# PRIMER OF PULPING AND PAPERMAKING

## Technologies and Production Practices

BY KEN L. PATRICK

Published by Miller Freeman Books
600 Harrison Street, San Francisco, CA 94107
Publishers of *Pulp & Paper* magazine

Miller Freeman
A United News & Media publication



Design and Typesetting: Jan Hughes

Library of Congress Cataloging-in-Publication Data
Patrick, Ken L.
Primer of pulping and paper making : technologies and production practices / Ken L. Patrick.
p. cm.
ISBN 0-87930-577-0
1. Wood-pulp. 2. Papermaking. I Title.
TS1175.P36 1999
676—dc21 98-51796
CIP

Printed in the United States of America
99 00 01 02 03 5 4 3 2 1

# Table of Contents

# Preface

THIS PRIMER IS DESIGNED FOR NON-TECHNICAL managers in the paper industry, financial and legal professionals, investment counselors, accountants, public and government relations supervisors, personnel directors, teachers, students, and others interested in knowing more about how pulp and paper are made. In easy-to-follow, non-technical terms, it covers current chemical and mechanical pulping processes and describes the latest paper machine systems and operations.

It should be helpful also to production and technical personnel within the industry who have concentrated most of their knowledge and experience in just one or a few process areas. By covering the integrated pulping and papermaking process in one quick-read, this primer provides a general overview of the industry and its various manufacturing technologies.

In the past, the typically large pulp and paper mill tended to operate as a cluster of separate plants and processing arenas. The woodyard took care of acquiring and processing wood. Its interests and concerns ended at the conveyor leading to the top of the digester.

The same was true of the pulp mill. Who knew or really even cared what happened to bleached and unbleached pulp once it left the high density storage tanks? And the paper mill was too busy keeping the paper machine up and running to nose around in finishing, converting, and shipping areas. And finishing had no real idea about what went on in the woodyard.

But today the focus has broadened considerably in all areas of a mill. With advanced process control and computerization came a slow but steady data interfacing and an emerging view of the operation as a single enterprise. Modern mills now know very well what goes on in all of the other sections and departments—on a real-time basis.

The problems encountered on a paper machine can now be easily and instantly tracked, forward and backward in the process, in search of causes, effects, and solutions. Information sharing on a major scale has just begun in the paper industry, but its potential is being sensed and examined at almost every level, from the mill operating floor to the corporate boardroom.

To better understand and interpret data and information flowing from and into modern pulp and paper mills, a working knowledge about all of the basic processes and manufacturing technologies will become increasingly critical. It is toward this end that this primer was developed.

# Into the Twenty-First Century

IN MANY WAYS THE GLOBAL PULP AND PAPER industry is rapidly becoming a distant relative to its predecessor industry of only a quarter century ago. Today's big, new, or modernized mills are largely computer controlled and more environmentally sound than mills of the 1970s and particularly mills operating in the first half of the twentieth century.

As a whole, the pulp and paper industry enters the twenty-first century in fairly good condition, at least process-wise, considering the literal evolution it pushed and pulled itself through in the 1980s and 1990s. In the environmental area, the industry had to generally abandon the use of elemental chlorine in its fiber bleaching processes, an established, highly effective, and economical chemistry in use by the industry for more than a hundred years.

In the wake of the dioxin scare in the mid-1980s, the industry converted practically overnight to various combinations of chlorine dioxide (a more environmentally friendly chemical cousin of elemental chlorine), oxygen, hydrogen peroxide, ozone, and other peroxygens. These ECF (elemental chlorine free) pulping and bleaching processes eliminated or dramatically reduced carcinogenic dioxin to non-detect levels and reduced other chlorinated organics in bleached pulp mill effluent streams. Some mills now operate totally chlorine free (TCF), using no chlorine-containing compounds at all in their bleach plants.

The industry also made considerable progress toward control of air emissions and solid wastes during the past 25 years. When the U.S. EPA's long-anticipated environmental Cluster Rules were finally promulgated in the late 1990s, most pulp and paper mills in North America were more or less already in compliance. For that matter, many modern chemical pulp mills around the world could have been considered in compliance with the Cluster Rules as well as environmental regulations in their own countries, particularly mills in Scandinavia.

But recent evolutions in the industry were not just environmentally oriented. Quality became a progressively important issue in marketplaces around the world, and ISO 9000 procedures and standards were rapidly adopted by almost every pulp and paper company and the large vendor-supplier sector surrounding the industry.

Process consistency, vital to quality and production economics, received a major boost in the late 1990s with the implementation of new on line sensor and control technologies. Maturing as the century closes, these advanced process control and information technologies have taken the industry to the threshold of complete automation. Another major contributor to improved process and product consistency, as well as process economics, has been and will continue to be advanced forestry genetics.

Field bus is currently approaching implementation in pulp and paper, as well as other industries. This new digital, standardized technology offers advanced control capabilities together with dramatically reduced installation costs for new sensors and control devices. The conversion to field bus is expected to slowly proceed during the next 10 to 15 years.

Progress toward process automation and advances in the environmental areas have moved pulp and paper mills around the world closer to minimum impact or effluent closure. Several mills in Europe, Scandinavia in particular, are now operating with only small amounts of effluent discharge from their processes.

Some mills, in fact, claim to have totally closed effluent systems. In North America and elsewhere in the world, significant progress has also been made toward the "closed mill concept." This trend will undoubtedly continue well into the twenty-first century.

Significant developments have occurred in the papermaking area as well. Paper machines have become progressively wider and faster during the past quarter century, and the introduction of gap formers and multi-wire front ends, wide-nip (shoe-type) presses, soft nip calendering, multiple coating, advanced minerals/microparticle technologies, improved drying techniques, etc., have elevated papermaking to a new level of art and science.

These newer pulping and papermaking technologies are discussed in the process sections that follow.

# Pulp and Paper Industry Structure

THROUGH MOST OF THE TWENTIETH CENTURY, the modern pulp and paper industry was concentrated mainly in Europe/Scandinavia, North America, and Japan. But in the latter decades of the twentieth century, big, new, modern pulp and paper mills sprang up in South America and more recently in various countries of Asia.

Expansions in South America and Asia have been primarily fueled by the development and growth of tropical plantation wood, primarily hardwood, but also softwood in some areas. Most of the industry's new or greenfield capacity is expected to continue tracking the emergence of these new "fiber bread baskets" into the twenty-first century.

The majority of the world's pulp and paper production is currently located in the U.S., which has a combined capacity equaling or exceeding the total of the next four largest producing countries. Recent figures from the American Forest and Paper Assn. (AF&PA) show that the U.S. represents only 5% of the world's population but accounts for 30% of the world's paper and paperboard capacity and some 34% of its pulp capacity.

The U.S. has the world's largest annual per capita consumption of paper and board products at about 740 lb per person, followed by Finland at around 670 lb, Belgium at some 565 lb, Japan at 560 lb, and Canada at 506 lb. Germany and the U.K. are at about 430 lb of per capita consumption and France is around 360 lb. By comparison, Malaysia's per capita consumption is approximately 200 lb, which represents a significant increase in recent years, following a trend in other areas of Asia. Mexico is less than half that.

During the late 1980s and 1990s, Asia, including China and Japan, was certainly a rising star in production of paper, paperboard, and pulp. This area currently represents around 30% of the world's total paper and board output, and should overtake Europe in the near future. Most of the new Asian tonnage is aimed at a rapidly rising domestic demand and replacement of import tonnages. In recent years, Asian producers have also increased their export shipments, with some of their paper products beginning to show up even in North America.

Although Japan and China still represent the majority of tonnages in Asia, Indonesia and Thailand have shown strong growth in recent years. In fact, Thailand's production now exceeds that of China. But China recently overtook Japan as the world's third largest pulp producer. Asia's share of the world pulp market is currently around 20%.

In the meantime, South America has continued its steady presence on the world markets. In Brazil, especially, very large market pulp mills have been started up and expanded during the past few decades, and some of these have now integrated with fine paper and board production capabilities, with a market focus toward Europe and North America as well.

In South America and Asia, tropical cloned hardwood plantations can yield mature trees from seedlings in six or seven years. In North America, by comparison, native hardwood maturity can take five to ten times longer. In both regions, the species are cloned not only for fast growth, but for disease resistance and superior pulp and paper qualities as well. Cloned plantation fiber is considerably less costly for these mills than fiber harvested from native and managed forests, and also results in a highly consistent fiber supply to a mill's processes and, thus, a consistently high-quality finished product.

Leading into the twenty-first century, U.S. paper and board annual capacity stood at about 90 million metric tons. This is followed by the total for Western Europe at around 82 million metric tons, total Asia (including Japan and China) at some 78 million metric tons, Canada and Eastern Europe at near 20 million metric tons each, and Latin America approaching 17 million metric tons (with Brazil accounting for about 45% of that).

The breakout for woodpulp is similar, with the U.S. having an annual capacity of around 63 million metric tons, followed by Western Europe with 39–40 million metric tons, Asia nearing 40 million metric tons, Canada holding at about 28 million metric tons, Eastern Europe at about 16 million metric tons, and Latin America at 12–13 million metric tons (more than half being in Brazil).

North America has about 172 of the world's 325 or so kraft mills, and employs more than 600,000 people overall. In the U.S. there are some 550 total paper and board mills. More than half of these produce less than 100,000 tpy, and more than 70% are below the 200,000-tpy threshold. Altogether, about 1,200 paper and board machines are operating in the U.S. There are some 250 pulp mills in the U.S. and a high percentage of these are integrated operations, i.e., they are located on the same sites and produce captive pulp for many of the 535 paper and board mills.

# The Science of Papermaking

THE ORIGINS OF MODERN PAPERMAKING can be found more in the arts than in the sciences. "Sheets" of paper made from the papyrus plant in Egypt date back as far as 3000 B.C. In fact, the name "paper" is derived from the Egyptian word "papyrus," a weed-like plant still growing along the Nile River.

Ancient Egyptians converted the papyrus leaf into a sheet by cutting it into thin slices, and then pressing them together. The use of papyrus spread out of Egypt and remained the major medium for written communications until the twelfth century A.D.

Another medium that developed a couple of milleniums after papyrus was parchment, which was made from animal skins. This practice came into common use sometime around 500 B.C., although there is evidence that the Hebrews wrote on animal skins as early as 1085 B.C.

Early parchment was made by steeping the skins of sheep and goats in pits impregnated with lime, and then stretching them on frames. These skins were then thinned by paring and scraping with sharp instruments. Vellum, a type of parchment, was made in a similar manner from the skins of young calves or stillborn lambs.

The beauty of these luxurious animal-skin writing materials is still preferred today. Simulated parchment is now made by passing webs of paper through acid baths. Modern vellum paper

*Making parchment.*

*Papermaking in 1568: The pulp was prepared by beating rags with wooden hammers, usually powered by waterwheels, visible on the upper left.*

is typically made with a high rag content and processed through supercalenders (special multi-roll presses) to resemble vellum.

The invention of paper as we know it today is credited to Ts'ai-Lun, a Chinese official in the court of Ho Ti, emperor of Cathay around 105 A.D. It was first produced at Leiyang, China, from fermented and beaten mulberry bark fibers. The fibers were suspended in water and scooped out with a screen made of bamboo fibers tied with horse hair.

*Hollander beater.*

*Hollander beater, 1927. These batch methods of pulp preparation have been replaced by continuous flow systems such as refiners.*

This method of making paper was kept secret from the rest of the world for several hundred years until the Arab armies captured Samarkand, a city in Western China, around 704 A.D. The Arabs learned the art of papermaking, a semi-science by this time, from captured Chinese papermakers who were taken to Baghdad to make paper.

It was at the end of this century that these early papermakers began to use linen rags as a primary source of fiber. The science spread to all Arab domains and passed into Europe with the invasion of the Muslims. The first paper made in Europe was at a mill in Toledo, Spain, in 1085 A.D. The Crusaders then spread the science of papermaking following their visits to Palestine and Syria in the twelfth century.

Papermaking appeared in France during the late 1100s and in Italy in the early 1200s, where watermarking first appeared on paper. Paper mills sprang up in Germany as early as 1336, and spread through Austria in 1356, Holland in 1428, Switzerland in 1433, Russia in 1576, and Scandinavia in 1630. Still made generally by hand, paper produced with textile and various vegetable fibers slowly replaced parchment and vellum during the fourteenth through the sixteenth centuries.

Papermaking arrived in North America in 1690 when William Rittenhouse and William Bradford built a mill at Wissahickon Creek in Germantown, Pennsylvania. The main fiber source for this mill was linen rags, a method that prevailed until the invention of the paper machine.

The first paper machine was invented by a Frenchman, Nicholas-Louis Robert, in 1799. His machine formed the fibers on a circular wire screen belt to produce paper in an endless sheet. However, Robert's invention was never really developed and commercially applied. The approach was continued by the English brothers Henry and Sealy Fourdrinier, who put a practical device into operation in 1810, giving birth to the now historically famous fourdrinier paper machine.

*This fourdrinier was state-of-the-art in 1927, producing 234"-width newsprint.*

Several years before the first fourdrinier type of paper machine started up, a cylinder type of paper machine was making its debut in England. In 1805, Joseph Brannah brought the cylinder concept into operation, but it was actually patented by another Englishman, John Dickinson, who consequently received credit for the invention.

The cylinder machine consisted of a rotating, screen-covered cylinder partially submerged in a vat of pulp. As the cylinder rotated, it picked up fibers from the vat, forming a continuous sheet of paper that was subsequently pressed and dried. The first paper machine in North America was a cylinder type of design, starting up in 1817 in Philadelphia.

Today's fast, wide paper machines have come a long way from these early fourdriner and cylinder machines, but the kinship ties remain no matter how far the technology seems to evolve. Modern, multi-wire and gap-type formers still employ the concept of endless sheet formation on (or between) a con-

*A very small amount of paper has continued to be made by hand, usually for artistic or decorative use.*

stantly moving screen-type of belt, now referred to as the forming fabric.

Modern paper machines operate at very high speeds to produce a myriad of papers and boards. Some of these machines are several stories high and can stretch for a city block or longer. They typically operate around-the-clock, seven days a week, being shut down for maintenance only once or twice a year.

The following sections describe the modern science of papermaking in basic, simplified terms. The discussion begins with preparation of the raw materials (wood and nonwood or agricultural fibers, chemicals, and water/steam) and proceeds through the various chemical and mechanical pulping technologies into contemporary methods of paper manufacture on various types of paper and board machines. Recycled fiber can

*Powell River BC was the largest single pulp and paper unit on the West Coast in 1927. The logs were brought down the river in rafts and stored in the log pond shown at the the front of the mill.*

also be considered a raw material and this technology is discussed in a separate section.

It should also be pointed out here that some mills are not integrated, i.e., they do not manufacture pulp on site for use in their paper or board machines. These mills either acquire pulp from another of their own company's mills or buy it from other "market pulp" producers. Conversely, other mills are not integrated with paper or board machines, producing pulp for use within their own company's structure or for sale on the open market.

Pulp and paper is one of the most-capital intensive industries in the world, due to the size of the equipment and systems used and the overall "massiveness" of the process. Depending on the grades being produced, the process utilized, and various construction factors, the cost of a modern chemical pulp and paper mill will fall in the range of $700,000 per daily ton of capacity.

Thus, a 1,000-tpd mill will cost nearly three-quarters of a billion dollars. Several recent greenfield mills have cost well over a billion dollars, and a few have approached two billion dollars. Building large mills takes advantage of economies of scale, but the extremely high costs of these green field ventures have also spawned a simultaneous trend toward smaller, recycled-based "mini-mills" in recent years.

# Raw Material Preparation

WOOD FIBER IS PROCESSED USING TREES harvested from forests or plantation growths. Nonwood or agricultural fibers used in pulping include straws, bamboo, bagasse (fiber mashings from sugar cane), reeds, rice, cotton, kenaf, hemp, etc. Very little nonwood fiber is used in the manufacture of pulp in North America or in Europe, although its use could increase in the not-too-distant future.

The use of wood as the major raw material for pulping and papermaking is expected to continue at least for the next quarter century. For that reason, this book's discussion of basic paper industry process technologies is focused primarily on this raw material.

Mills typically purchase chips or process their own chips on site from whole logs (roundwood) they buy or harvest from their own woodlands operations. Some mills that use a mechanical rather than a chemical pulping process (or a combination of both) might use roundwood without chipping, or they could use a combination of both chips and roundwood (see section on mechanical pulping below).

Purchased chips are received by truck, rail, or barge and stored in outside chip piles or chip silos, depending on the wood species. Wood converted into chips on site in the mill's woodyard goes through several stages of processing. Logs of various lengths are delivered to the woodyard and stored in well-managed piles. "Shortwood" logs which are cut to length (usually refers to less than 120 in.) in the woods before being delivered to the mill. "Longwood," or tree-length/random-length logs, became the preferred form several decades ago, because less fiber is lost or left on the forest floor if the trees are not cut into shortwood.

Some companies maintain "satellite" woodyards near the harvest points where trees are collected and further processed for shipping to the mill. These satellite operations supply either whole or chipped wood to the mills. Today, most wood used by the industry is purchased from pulpwood dealers (who usually buy from private landowners) or from sawmills in chip form.

In the mill, roundwood is processed either as shortwood or longwood, depending on the mill's specific process. Some mills cut the longwood into shorter lengths across massive slasher or saw decks, but the processing of tree-length wood has come into wider practice in recent years because it can be more efficient and can maximize chip yield.

Before being chipped, the logs are debarked in several ways. The most common method used in most modern mills employs giant revolving "barking" drums. In these units, bark is stripped from the logs by friction as they tumble against each other and the steel-chambered wall of the drum. Bark can also be removed by moving the logs past streams of high-pressure water in a hydraulic barker, or it can be removed by sets of mechanical knives.

The debarked logs then go either directly to a mechanical pulping process or they are chipped for use in chemical pulping and some mechanical pulping processes. In the chippers, logs are dropped by gravity against a rapidly revolving disc with heavy-duty, sharp knives set at an angle to produce chips approximately one-half to three-quarters of an inch in size.

*These chips are being conveyed to a chip screen.*

These chips and any unscreened chips from the mill's woodlands operations pass over vibrating screens. Oversized chips are removed and sent to a rechipper for further size reduction. The rechipped "overs" are then sent for another pass through the vibrating screens. Typically, fines and sawdust, along with bark removed in the earlier stages, are sent to the mill's power boiler for fuel (in mills that have the capacity for burning wood residues and biomass).

The chips pass through various other stages of screening, with final accepts being transported to either the outside chip storage piles or silos as mentioned above. The chips are generally retrieved on a first-in/first-out (FIFO) basis and sent to smaller tanks or silos in the pulp mill before being metered into the mill's digesters or cooking vessels.

# Chemical Pulping

WOOD COMPRISES 70% TO 80% CELLULOSE and hemicellulose fibers. The remainder is lignin (a brown, rigid, glue-like substance that binds the woody fibers together), sugars, gums, resins, and mineral salts. In the pulp mill, the cellulose fibers are separated from each other as well as the other "impurities" by either chemical (cooking) or mechanical means, or a modified combination of both technologies (semichemical).

Chemical pulping does a more thorough job of softening and removing the lignin and other impurities, while mechanical processes leave a high percentage of lignin in the pulp, resulting in a much higher "yield." Chemical processes typically provide a pulp that is stronger (separated fibers interlock in the paper forming process) and more easily bleached.

Thus, pulps from the chemical processes are usually used in board grades where strength is critical and in fine paper and packaging grades where both strength and appearance (brightness in particular) are important characteristics. Mechanical pulping is generally less expensive per ton of production (and in capital investment terms) than chemical pulping, and produces a high-yield pulp with high opacity and bulk suitable for use in newsprint and other specialty grades where these characteristics are important.

However, recent advances have produced a "marriage" of sorts between traditional chemical and mechanical pulping

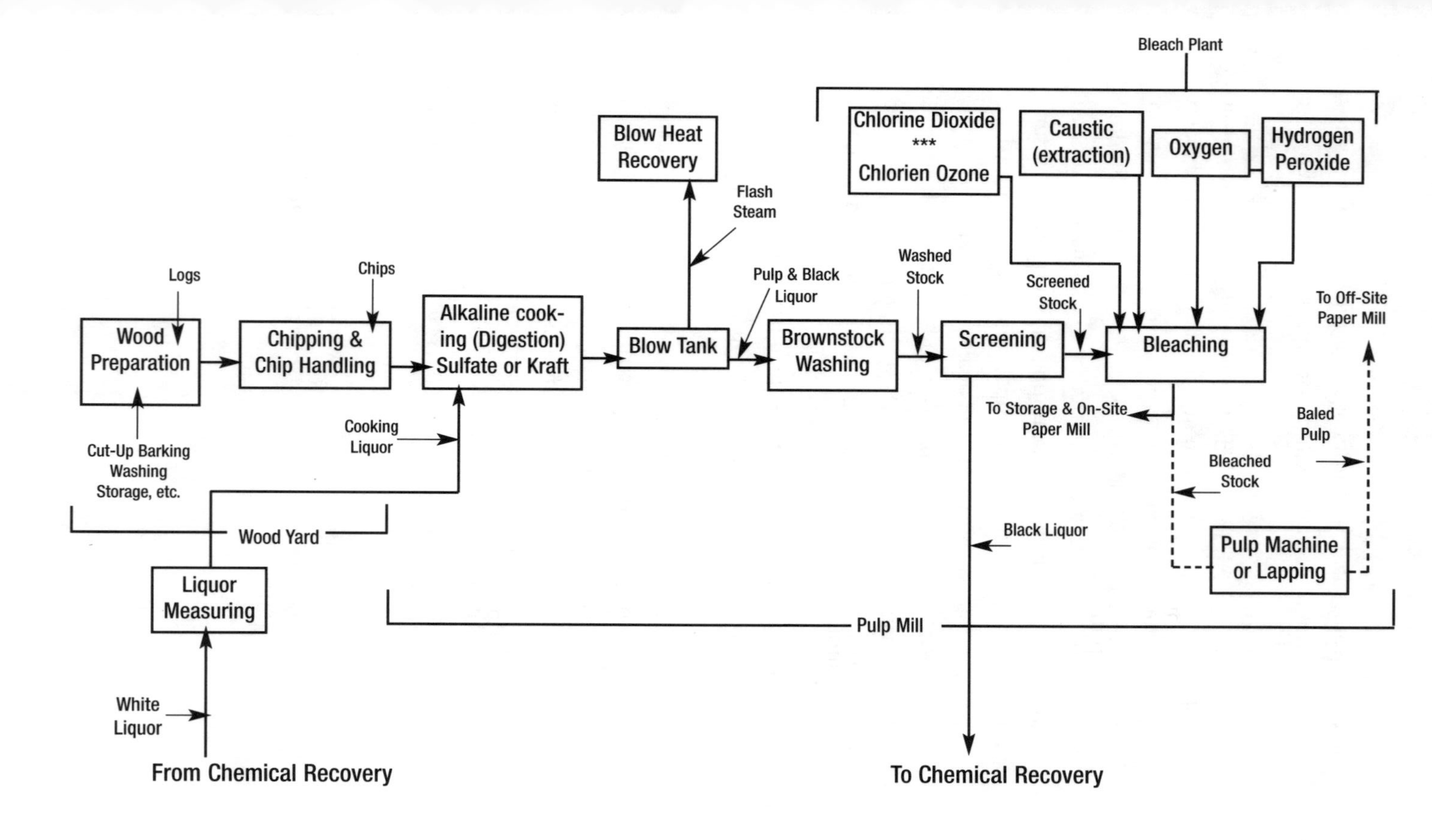


*Figure 1: Block diagram of the sulfate or Kraft pulping process.*

processes. Some of these pulps, such as bleached chemithermomechanical pulps (BCTMP), combine several technologies and preserve some of the key characteristics of both chemical and mechanical processes.

There are basically three categories of chemical pulping—sulfate (kraft), sulfite, and semichemical or NSSC (neutral sulfite semichemical), as it was originally called. Today, in North America especially, the sulfate or kraft process predominates. Some 70% of North American pulp production is now done with the kraft process.

SULFATE OR KRAFT PROCESS. The sulfate process was invented by Carl F. Dahl in Germany around 1884. In German, the word "kraft" means "strength" or "strong." With this process, practically any species of wood can be used. The process produces a strong, easily bleached fiber that is used in many grades of paper and board. In printing and writing papers, bleached softwood kraft (BSK) is typically blended with bleached hardwood kraft (BHK) fibers to produce a sheet with excellent brightness, surface smoothness, opacity, and printability.

BSK is also a major component in various packaging grades, newsprint (blended with mechanical pulps and recycled fibers), and tissue. Unbleached kraft (UBK) is typically the key component in bag papers, shipping sacks, wrapping papers, and containerboards used in the manufacture of corrugated shipping boxes.

The kraft process (depicted in Figure 1) is also economically attractive because the spent cooking liquor from the process (black liquor) can be recovered, concentrated, and burned in a mill's recovery boiler. This process allows regeneration of the cooking chemicals, and produces steam for cogeneration at the same time (see section on chemical recovery below). The disadvantages of the kraft process include high capital costs, relatively low yields (45%–55%), corrosiveness to metals and equipment, and environmental challenges.

The characteristic odor accompanying the kraft process is due mainly to methyl mercaptans and sulfur compounds liberated during the cooking or digestion stage. While not necessarily an environmental threat, these odors can be quite offensive. However, major strides have been taken toward odor control of the kraft mill in recent decades.

Kraft or alkaline pulping actually has its roots in the soda process of the 1880s, in which cooking was carried out in a solution of caustic soda (sodium hydroxide) at a pH around 12. However, acid generated during the cook tended to lower the efficiency of this process by reducing the pH in the digester. Later, sulfide was added to "buffer" the cook, resulting in the basic kraft process used today. The key chemical components in modern kraft processes are still sodium hydroxide and sodium sulfide, known as "white liquor."

Kraft cooking is carried out in large batch or continuous vessels called digesters. In the batch process, chips and white liquor are metered into the top of the digester, which is then sealed, pressurized, and heated by the addition of steam. After several hours (eight or more hours is common), with pressures of 90–120 psi and temperatures rising to around 350°F, the digester is "blown" into a blow tank. The pressure of the discharge or "blow" against baffles in the blow tank helps break up the fiber bundles, which are still more or less in chip form, and it also aids in separating the spent cooking liquor (called black liquor because of its color) from the fibers.

In the continuous digester, the chips are presteamed in the presence of some white liquor. They are then fed continuously into the cooking vessel. In the cooking zone, hot white liquor is pushed through the chip mass from a pipe running down the center of the unit. As the chips pass down the vessel (via a screw-type conveying apparatus), washwater and/or spent cooking liquor is forced through the chip mass. After a final washing zone, the cooked chips are drawn out the bottom.

*This continuous digester is in use at Weyerhaeuser Co.'s Flint River fluff pulp mill in Oglethorpe, GA.*

Modifications to the continuous process in recent years have improved overall cooking efficiency and helped reduce subsequent bleaching demand, with significant environmental and quality benefits.

These "extended delignification" modifications are marketed under various acronyms, such as MCC (modified continuous cooking) and EMCC (extended modified continuous cooking), as supplied by Ahlstrom (Kamyr) Corp. The MCC process uses cooking in the vessel's traditional upper portion (concurrent) followed by a second (countercurrent) cooking zone. With EMCC, the chips are "cooked" in yet another zone, the traditional high-heat wash zone, where fresh white liquor is added and upflows through the chip mass. Kvaerner Pulping Technologies' similar approach to extended delignification in a continuous vessel is called Isothermal Cooking (ITC).

Extended delignification technologies have also been developed for batch digester systems. Beloit, for example, uses a series of special tanks to "shuttle" various liquors and washes in and out of the batch digester in its RDH (rapid displacement heating) system. A similar approach was used by Voest-Alpine in its Enerbatch system, which was acquired by the Impco Div. of Ingersol Rand, which subsequently was acquired by Beloit's Fiber Systems Div. in the mid-1990s. The Sunds Defibrator extended delignification system for batch digesters is known as Superbatch.

These modified systems have the effect of a "gentler" cook. Repeated applications of cooking liquor over extended periods of time result in a stronger and better delignified pulp compared with that produced in one harsh "shock" application.

After being discharged from either a batch or continuous digester, the pulp (now known as "brownstock" because of its characteristic color at this stage) is forced through screens to strain out knots and incompletely-cooked chips. The brown stock is then washed in a series of brown stock washers, often followed by a series of pulp mill refiners (sealed units with large rotating discs) that further shape and prepare the pulp.

In recent years, mills producing bleached pulps, especially, have added an additional stage of "delignification" beyond the digester, in which the pulp is exposed to fairly high dosages of oxygen in large tanks known as reaction vessels. "Oxygen delignification," as the stage is called, can be done in one or two vessels (with an intermediate washing stage).

The main objective of oxygen delignification in recent years has been to further reduce the "kappa number" of cooked pulps, which is a standardized measure of the pulp's general bleachability in subsequent stages. Some oxygen delignification stages can reduce kappa number 40%–50% or more.

In addition to extended delignification modifications and oxygen delignification stages, several digester chemical additives have been used to improve the efficiency of the cook and further reduce kappa number without adversely affecting quality, and in some cases to boost yield at the same time. Additives such as anthraquinone (AQ) and polysulfide (PS), for example, have been used in this regard, with sporadic success. In most cases, costs of these additives have been a limiting factor, but they could play an increasingly important role in the twenty-first century.

Also, the use of enzymes and various fungi-based "bio-pulping" techniques were examined and tried during the 1980s and 1990s, with mixed results. These environmentally-driven technologies will likely find renewed application in the next quarter century.

Black liquor extracted from the pulp at the brown stock washers (called filtrate) is sent, along with that recovered in the blow tank and oxygen delignification washer filtrate, to the chemical recovery process. As described below, the collected black liquor is concentrated in evaporator trains and burned in large recovery boilers, producing energy and a chemical smelt reprocessable almost endlessly into kraft cooking chemicals.

The screened, washed, and refined brown stock is sent to large storage vessels called brown stock storage tanks, or, because the pulp has usually been adjusted to a high consistency, these are also referred to as brown stock high density storage tanks. From high density storage, the brown stock goes either directly to the paper machine area for production of unbleached paper and board products or to the bleach plant for further removal of lignin and impurities and subsequent

*This bleach plant upflow tower is at the Bahia Sul Celulose Sa mill in Brazil.*

use in printing and writing papers, bleached boards, packaging grades, etc.

THE KRAFT BLEACH PLANT. The main bleaching agent for kraft pulps today is chlorine dioxide (D), although elemental chlorine (C) is still used in various parts of the world, including North America. In the bleach plant, the brown stock is mixed with the bleaching agent and pumped in stages through tall, cylindrical vessels called towers.

After the initial "chlorination stage" (if the mill is using a chlorine containing compound in the first stage), the stock passes through an "extraction stage" (E) in which sodium hydroxide (caustic) is usually used to remove the water insoluble chlorinated lignin and other colored compounds still remaining with the pulp.

The stock then passes through other bleaching stages and associated towers involving the application of more chlorine dioxide (D), oxygen (O), hydrogen peroxide (P), Ozone (Z), etc. Sodium hypochlorite (H) has been used in these sequences, but has fallen out of favor in recent years because of problems with chloroform emissions.

The order in which the bleaching agents are progressively applied to the stock is commonly referred to as the bleaching sequence. For example, the bleaching sequence DEDED would indicate three sequential stages of chlorine dioxide (D) separated by caustic extraction stages (E).

A $D_CE_{OP}DED$ sequence, as another example, begins with chlorine dioxide (primarily) and chlorine being used together in the chlorination stage ($D_C$), followed by subsequent stages of caustic extraction "reinforced" with the addition of oxygen and peroxide ($E_{OP}$), chlorine dioxide (D), caustic extraction (E), and a final chlorine dioxide stage (D). The use of chlorine dioxide to replace or substitute for elemental chlorine in the chlorination stage has been commonly referred to as "substitution."

The bleaching agent with the highest content percentage in the chlorination stage is generally shown first as a capital letter in the sequence, with the minor agent being either a lower case or subscript letter, as in the $D_C$ designation above. Mills that have completely substituted chlorine dioxide for elemental chlorine are said to have a $D_{100}$ process.

*The secondary effluent treatment lagoon at Bahia Sul, Brazil, contains 80 aeration flotation cells. Sludge is landfilled or used as a soil fertilizer.*

A typical non-chlorine bleaching sequence might by $OZE_OP$, where oxygen (or the oxygen delignification stage) is considered to be the initial stage (O), followed by treatment in a special ozone reactor tower (Z), oxidative reinforced extraction ($E_O$), and a hydrogen peroxide stage (P).

Ozone is being used increasingly in kraft bleaching sequences, especially in Scandinavia, but it must be used judiciously and carefully since it tends to be a "mad dog" agent, attacking cellulose fibers and lignin with equal vigor. Other peroxygens such as peracetic acid and caros' acid have also been investigated as potential bleaching agents, but transportation problems with these chemicals have not yet been worked out.

Bleach plant washers, similar to brown stock washers in the pulp mill, are used in and between some of the bleaching stages. The main objective of many, if not most, of the oxygen, ozone, and peroxygen applications in pulp bleaching have been aimed at reducing chlorinated organics in the bleach washer filtrates, especially the most potent carcinogenic "chlorotoxins" such as dioxins and furans. The chlorinated organic content in a mill's effluent stream is commonly measured as AOX (absorbable organic halides), a test that produces a single value for all chlorinated organics in the effluent discharge.

Bleached pulp is stored in high density storage tanks before being sent to the paper machine stock preparation area. The brightness of bleached pulp is usually measured as ISO brightness or GE brightness, depending on the instrumentation and calibration used to measure it.

Generally, anything above 90% GE or ISO brightness is considered "market brightness" and is suitable for making higher quality printing and writing papers and specialty packaging grades. Actually, fine paper brightness can range from the low- to mid-80s to the mid- to high-90s. Newsprint, by comparison, ranges from the low- to mid-50s to around 60 brightness.

THE KRAFT CHEMICAL RECOVERY PROCESS. Black liquor extracted from the digester blow and pulp mill washer filtrates is a complex mixture of various organic compounds and inorganic salts. As depicted in Figure 2, this "weak black liquor" (12%–16% solids) is collected in a tank and then sent to heated (steam or other sources) multiple-effect evaporator trains where the percent solids level is slowly increased and the density of the black liquor rises.

Different types of evaporators are used, including direct contact evaporators (usually heated by recovery furnace exhaust gases) of the cascade or cyclone design, and the falling film and rising film designs. The types of evaporators differ mainly in the way that the liquor is exposed to the heat-exchanging surfaces.

"Strong black liquor" from the evaporators (50%–55% solids) is sent to a strong (or thick) black liquor storage tank. It is then raised to higher solids in a special direct-contact type of evaporator known as a concentrator. Firing into the recovery boiler is done at solids levels above 60%. Some mills today, in fact, are firing at very high solids in the

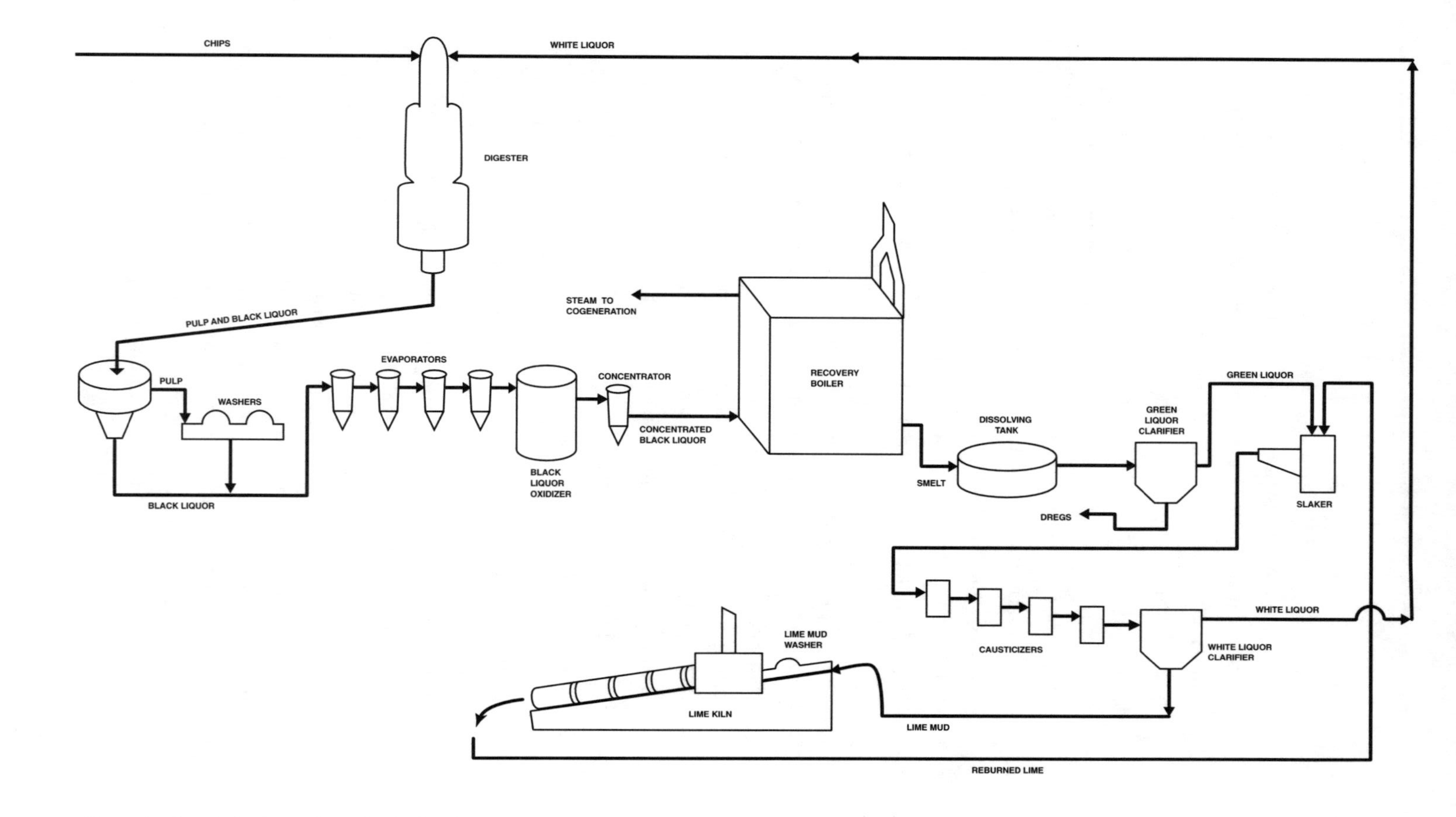


*Figure 2: Chemical recovery loop for Kraft pulping process.*

70%–80% range, with improvements in black liquor viscosity control.

The Black Liquor Recovery Boiler Advisory Committee (BLRBAC) requirements stipulate that an automatic system must divert liquor away from the boiler should levels drop below 58%. If solids concentrations are too low, the boiler is subject to dangerous steam explosions in the char bed area.

The black liquor (strong and/or weak) is usually sent through a stage known as black liquor oxidation. In this step, unstable sodium sulfide is transformed to the more stable sodium thiosulfate by the use of air or oxygen. This minimizes the formation (and loss) of objectionable hydrogen sulfide and organic sulfur compounds during evaporation and combustion in the recovery furnace.

The concentrated black liquor is sprayed into the recovery boiler via special liquor feed guns. These guns atomize the liquor into fine droplets. The organics in the liquor spray combust and provide heat for steam used in cogeneration (electricity and process steam), while the inorganics build up in a molten smelt bed on the floor of the furnace.

The molten smelt flows out of the recovery boiler through spouts into a vessel known as a dissolving tank. Here it is mixed with a weak wash taken from the lime mud washer or pressure filter further up in the recovery process. The mixture in this tank is called "green liquor" because of its characteristic color. The primary component of the recovery boiler smelt is sodium carbonate. The basic function of the chemical recovery or causticizing area is to convert this sodium carbonate into sodium hydroxide and sodium sulfide, the major components of white liquor (so called because of its characteristic clear-white color).

The green liquor mixture is pumped from the dissolving tank to a green liquor clarifier where various foreign materials or "dregs" are removed and usually sewered. The clarified green liquor is then sent to a slaker where it is slurried with reburned lime.

The lime (provided by an on site lime kiln) is "classified" to remove grit and poorly burned lime pellets, and carefully metered into the green liquor based on its exact sodium carbonate level. New on line sensor developments are helping to dramatically improve the "causticizing efficiency" of this operation.

The lime mud slurry is then pumped to the causticizers where the conversion of sodium carbonate to sodium hydroxide and sodium sulfide continues through several stages (typically three) of mixing, classifying, and washing.

This slurry emerging from the causticizers is sent to the white liquor clarifier where clear white liquor is separated or decanted from the mud and sent to the digester cooking chemical preparation area. The mud extracted from the bottom of the clarifier (primarily calcium carbonate at 35%–40% solids) is sent to the lime mud washer or pressure filter where it is increased to about 65% dry solids.

Recent improvements in filter design and the quality of the mud being filtered have allowed a dry solids feed as high as 80% into the kiln. The filtrate from the washer/filter at this stage is sent back to the dissolving tank as described above.

The lime mud is sent to a storage tank and fed into the long, cylindrical lime kiln. Under extremely high temperatures, the kiln converts the calcium carbonate component in the mud to calcium oxide, commonly known as lime. The "reburned" lime is then sent to the slaker for causticizing of the green liquor into white liquor as described above. Fresh lime makeup is used as needed to compensate for losses in the system.

SULFITE PROCESS. This process uses mainly softwood species such as spruce, hemlock, and fir. It was invented by Benjamin C. Tilghman in 1867. Sulfite pulping has some economic advantages and some quality and environmental trade-offs compared with the kraft process. Sulfite pulp is not as strong as kraft pulps, but generally it is much brighter using the same species and does not require as much bleaching, making it conducive to non-chlorine, TCF bleaching sequences.

Sulfite pulp is typically bulkier, softer, and more absorbent than kraft pulp, and thus is particularly suitable in tissue and sanitary grades, as well as some fine paper grades due to good surface formation and smoothness The sulfite process also does not produce as many offensive mercaptan and sulfur odor problems as kraft.

The sulfite process faded from popularity beginning in the 1930s for various reasons, but especially because of chemical recovery limitations. Currently there are only about 25 sulfite mills operating in North America, producing about 2 million metric tons per year. In Europe and the former USSR block countries, and elsewhere in the world, there are "scatterings" of sulfite mills still operating.

In Germany, for example, the kraft process was phased out in the recent past (reportedly for environmental reasons), but several aging sulfite mills are still producing in that country. Interestingly, Germany appears to be lifting its kraft "ban," and some new, environmentally-improved kraft mills will likely start up there in the early years of the twenty-first century.

Sulfite cooking is done under acid rather than alkaline conditions as with the kraft or sulfate process. Cooking liquor consists of sulfurous acid and a salt of this acid produced by burning sulfur to sulfur dioxide. The desired cooking liquor is formed by the absorption of the sulfur dioxide gas by the hydroxide of a base chemical. Formerly calcium was used as the base, but this approach prevented chemical recovery and was generally abandoned.

Newer sulfite mills use recoverable bases such as magnesium, sodium, and ammonium. Figure 3 depicts a typical

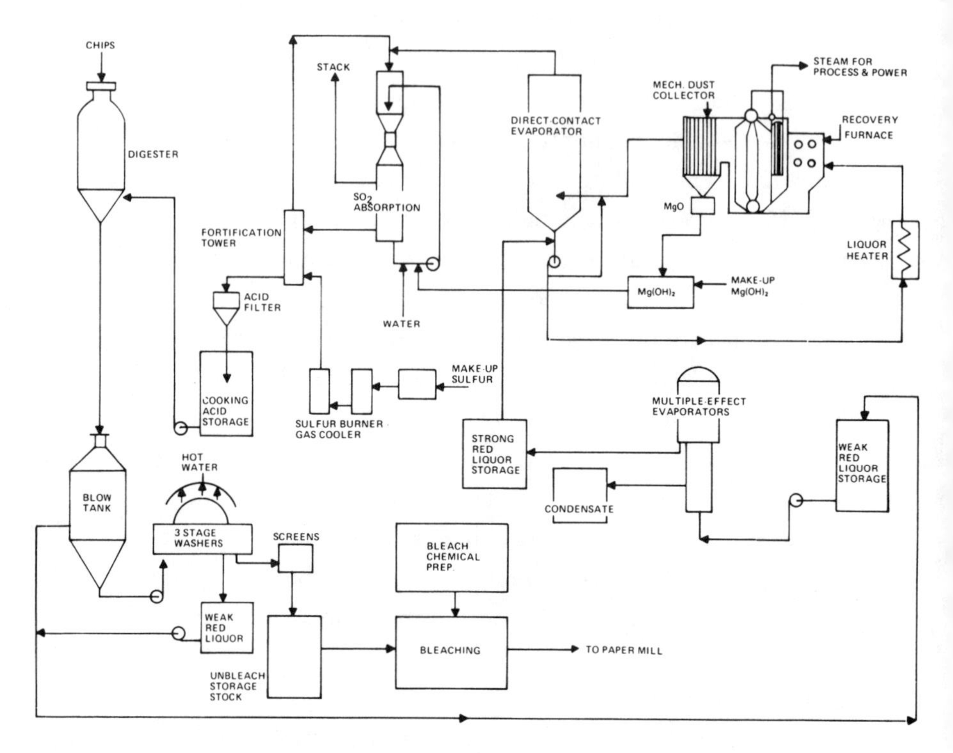


*Figure 3: Flow diagram of a typical magnesium-base sulfite pulping process.*

pulping and chemical recovery process for a magnesium-based sulfite pulp mill. Basically, this process resembles that used with the kraft process, except that the liquors are referred to as weak red (or pink) liquor and strong red (or just "red") liquor because of their characteristic colors.

The spent liquor and washer filtrates (pink or weak red liquor) in the magnesium-based recovery process are sent through multiple effect evaporators and transferred to the strong red liquor storage tank. They are then burned in a recovery boiler as with the kraft process. The combustion products of the sulfur and magnesium components in the liquor are discharged from the furnace in the gas stream as sulfur dioxide and solid particles of magnesium hydroxide.

The sulfur dioxide is recovered by reaction with the magnesium hydroxide to produce a magnesium bisulfite acid. This

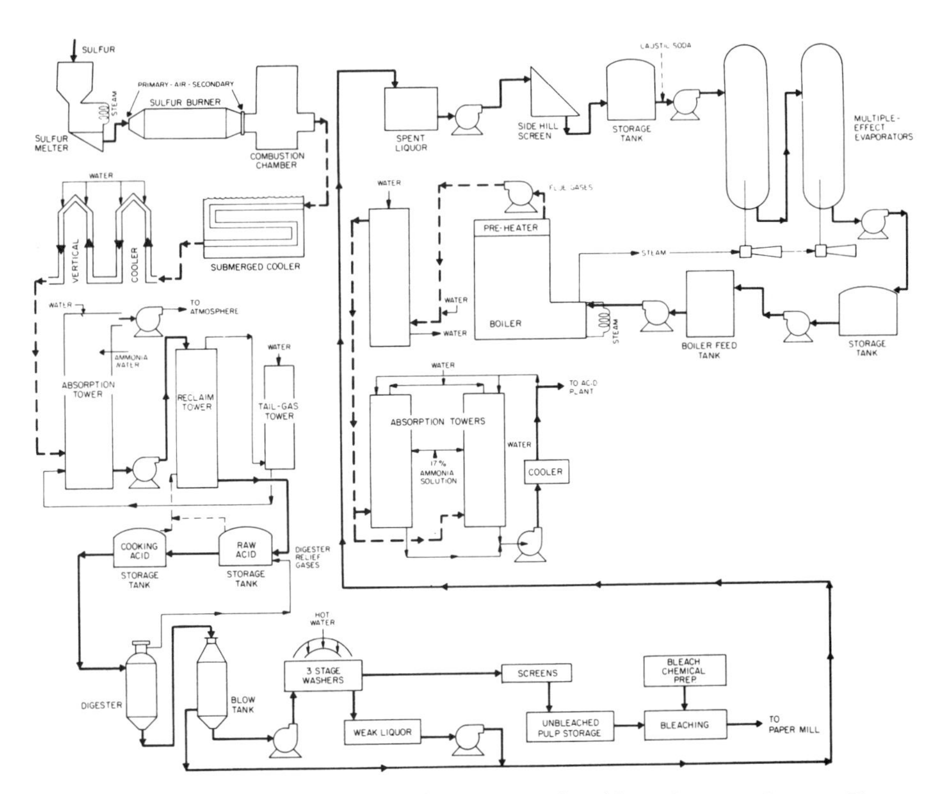


*Figure 4: Flow diagram of a typical ammonium-base sulfite pulping process.*

acid is then passed through a fortification, or "bisulfiting" system and fortified with makeup sulfur dioxide. The finished cooking acid is filtered and stored for reuse in the digester.

In the ammonium-based process, the liquor is burned for its heat value, but its ammonia values are not necessarily recovered. However, sulfur dioxide is recoverable from this process. Figure 4 shows a typical ammonium-based sulfite pulping and recovery process.

The newest sulfite technologies are much more environmentally sound and produce a higher quality pulp than their counterparts described above. Some of the new sulfite approaches are alkaline or neutral rather than acid-based. For example, aspen prepared using the NSAQ (neutral sulfite anthraquinone) process requires only two stages of bleaching with oxygen and peroxide to attain an 88% brightness. Yields with the aspen

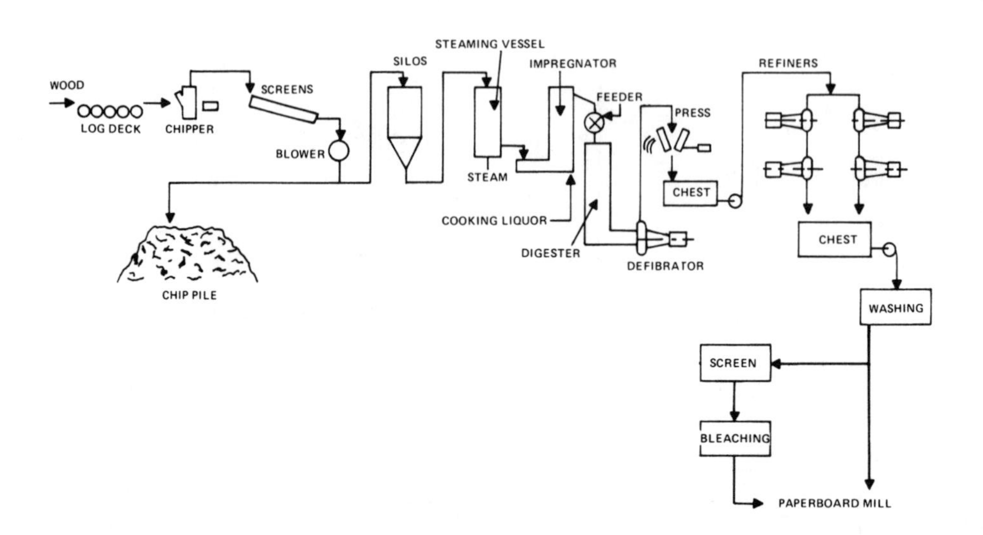


*Figure 5: Simplified flow diagram of a typical NSSC pulping process.*

NSAQ process are up around 70%, which is likely to be the upper ceiling with chemical pulping.

The extension of high-yield sulfite pulping, also known as CTMP (chemithermomechanical pulping), beyond the 80% yield range has now become a commercial reality. In this process, mechanical action (higher yield) is substituted judiciously for chemical action (lower yield). These processes are actually in the semichemical domain, on the borderline between chemical and mechanical processes.

SEMICHEMICAL PROCESS. Semichemical pulping was discovered in the late 1880s, but came into commercial practice in the 1920s. It was initially commercialized to take advantage of a relatively abundant hardwood fiber base that was not being used much at the time. However, semichemical pulping is sometimes used with softwood chips and sawdust as well.

Basically, this process combines chemical and mechanical methods. The wood is given just enough chemical treatment to soften the lignin and loosen the fibers. Separation of the fibers, or defibrating, is then accomplished by mechanical means. The semichemical process typically produces a stiff, resilient furnish used in such products as corrugating medium (used to make the corrugated inner lining of shipping boxes), egg cartons, etc.

There are basically three types of semichemical processes—neutral sulfite, cold soda, and chemigroundwood.

NEUTRAL SULFITE SEMICHEMICAL (NSSC). In this original semichemical process, hardwood chips are cooked in a sodium sulfite liquor at a neutral pH. Sodium carbonate, or soda ash, is added to buffer acid produced during the cook. Figure 5 depicts one of several NSSC processes that are used in the production of pulp for corrugating medium. Rotary cooking vessels were initially used, but various continuous vessels are used today.

Screened chips are heated in a steaming vessel and pass into the impregnation vessel by a screw conveyor, where they are exposed to cooking liquor. Times in the digester are relatively short, sometimes no more than 30 minutes at temperatures around 350°F. The semi-cooked pulp passes through a defibrator where the chips are broken up and sent through a screw or disc press.

After the press, which removes spent chemicals and organic matter for burning in a fluidized bed incinerator or other special unit, the coarse NSSC pulp is sent through refiners and subsequent washing and screening. The refined pulp then goes to the paperboard machine, typically unbleached. But NSSC pulps can be bleached.

In the late 1960s and early 1970s, a non- or no-sulfur semichemical process was commericalized that uses a solution of sodium hydroxide and soda ash to produce a pulp equivalent in quality to NSSC. It can be easily retrofitted to existing processes, and a number of NSSC producers have since converted to the no-sulfur, soda-based cook.

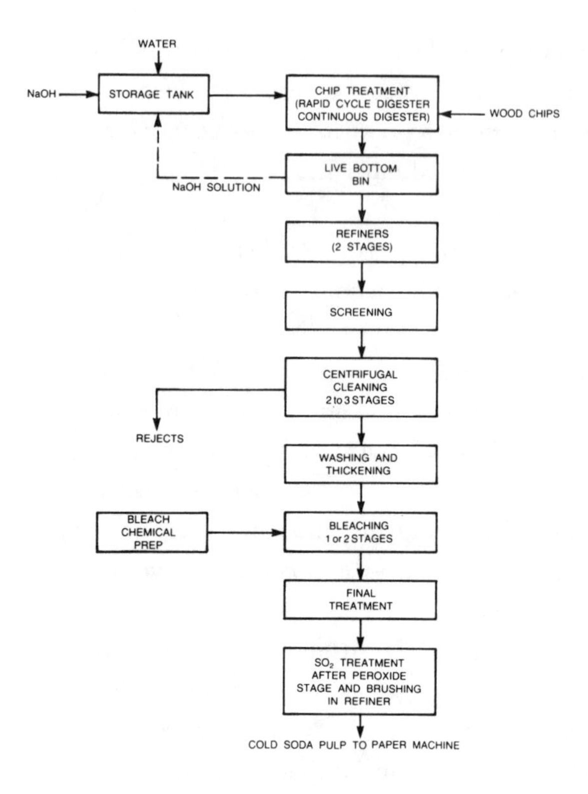


*Figure 6: Major steps in the cold soda pulping process.*

COLD SODA PROCESS. In the cold soda process, developed between 1919 and 1930, chips are treated with cold sodium hydroxide using hydrostatic, mechanical, or atmospheric pressures. Key stages in the cold soda process are shown in Figure 6.

After cooking, the chips pass through a live bottom bin into two or three stages of refining, screening, centrifugal cleaning (units that spin the stock at high speed, centrifugally separating contaminants from the fibers), washing, and thickening. If the stock is to be used in unbleached grades of board, it goes directly to the paper machine.

Cold soda pulp can be bleached and used in printing papers and boards. Typically, the bleaching is done at high consistencies in one to three short stages. The spent liquor is collected from the washing and pressing stages and fortified with fresh sodium hydroxide liquor for reuse in the process.

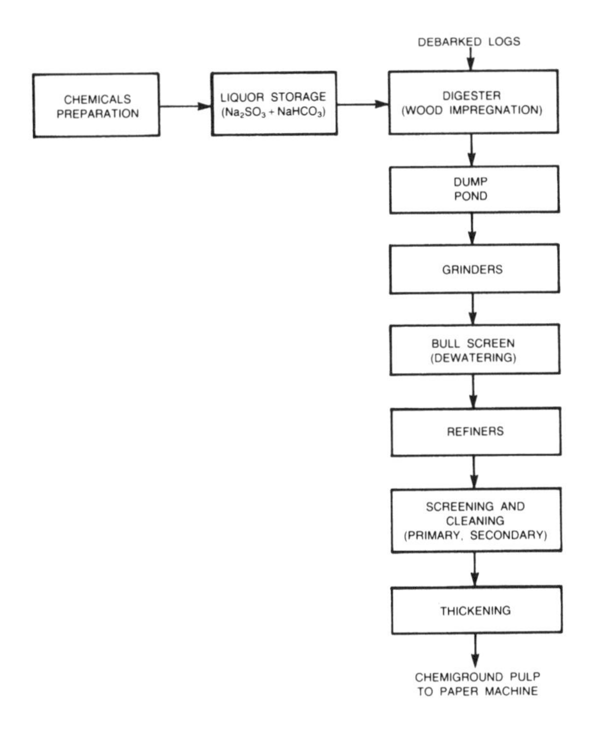


*Figure 7: Major steps in the chemigroundwood process.*

CHEMIGROUNDWOOD PROCESS. Introduced around the turn of the twentieth century, this process did not find commercial application until the 1940s. Basically, it involves chemical treatment of short logs (about 4 ft in length) prior to grinding them into fiber.

As depicted in Figure 7, the debarked logs are fed into a digester and cooked under pressure for a short period of time. The hot cooking liquor is a sodium sulfite/sodium bicarbonate solution.

After cooking, the logs are sent to a dump pond and subsequently withdrawn for defibering using conventional wood grinders. The stock is then refined and sent through a series of screens. After thickening, it is sent to the paper machine. As with cold soda, the spent chemicals are fortified with fresh sodium sulfite and sodium bicarbonate and reused in the process.

# Mechanical Pulping

INVENTED IN THE MID-1800S, mechanical pulping is the simplest form of pulping, although the processes became progressively advanced and sophisticated throughout the twentieth century. Essentially, in these processes the wood is reduced to its fiber components by mechanical force. As mentioned earlier, mechanical pulp fibers are ideal for grades where bulk and opacity are desired properties, such as newsprint and various groundwood specialty grades. In fact, some of the lightweight coated publication grades (No. 5 and some No. 4 in the U.S., for example) have a groundwood or mechanical fiber content.

But because of their low strength, grades made with mechanical pulps usually are blended with kraft or sulfite pulps for added strength on printing presses. There are basically three types of mechanical pulps—groundwood (GW) or stone groundwood (SGW), refiner mechanical pulp (RMP), and thermomechanical pulp (TMP).

There are further variations of these three types. For example, pressurized groundwood (PGW) and groundwood from chips (FGP, or fine groundwood pulp because of the fine fiber produced) are variations of the basic groundwood process. Chemi-thermomechanical pulp (CTMP) is a variation of thermomechanical pulping, and as mentioned above, is a borderline semichemical process also, especially the bleached (BCTMP) grades.

STONE GROUNDWOOD (SGW). In this process (simplified in

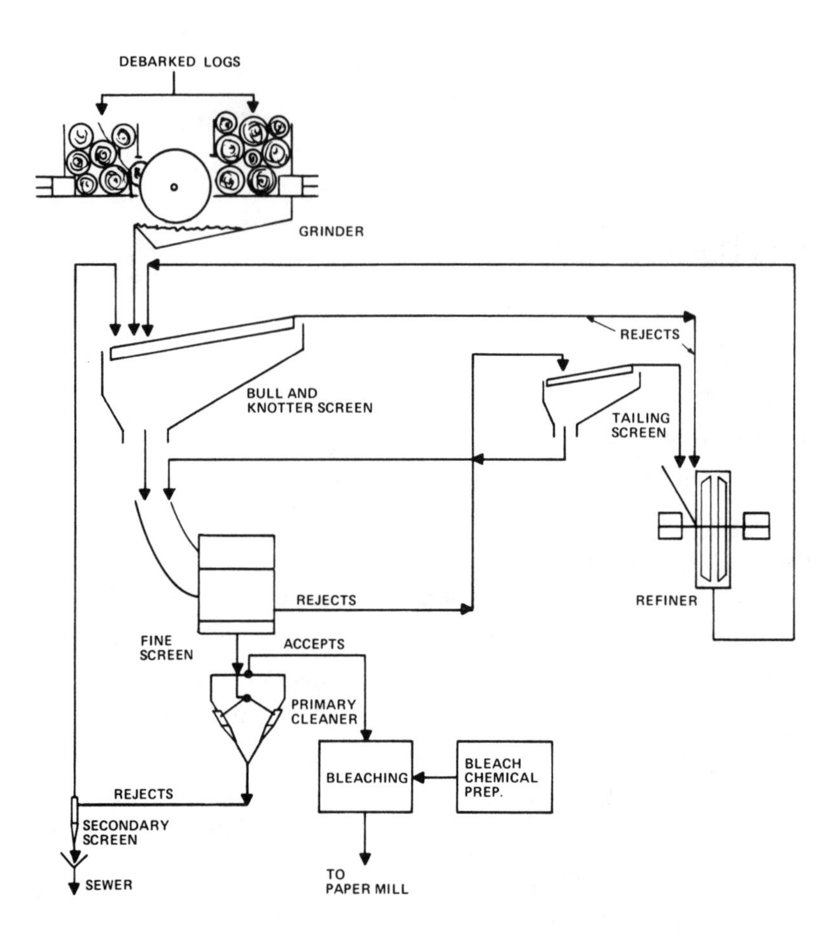


*Figure 8: Simplified flow diagram of a typical ground-wood mill.*

Figure 8), debarked short logs are pressed by gravity or hydraulic pistons through "pockets" against large, rough grinding stones that today are generally driven by big electric motors. The frictional heat generated in the grinding process tends to "plasticize" the lignin around the fibers, which helps to protect them from further damage.

The ground fibers are washed away from the grinding stone into a chest from which they are screened (several stages ), with rejects being sent through refiners to break up knots, slivers, and shives (oversized pieces). These rejects pass back through

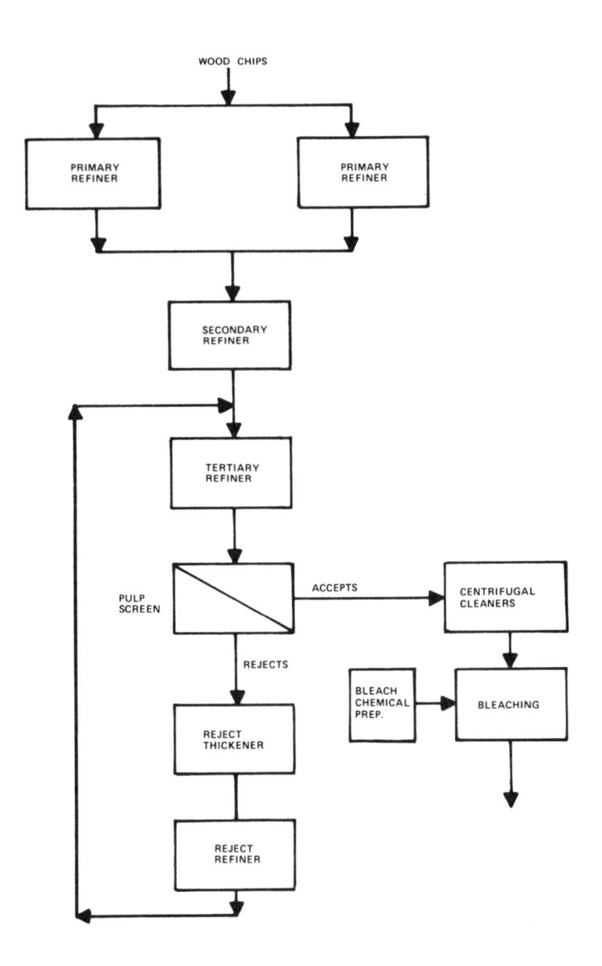


*Figure 9: Block diagram of a typical refiner mechanical pulping process.*

the screening system. Accepts from the screening system pass through centrifugal cleaners and go to the paper machine as either bleached or unbleached furnish.

Although this process is very energy intensive, the pulp yield is high at around 90%. Also, because the lignin is not removed, brightness of these grades is not nearly as high as that of chemical pulps, and the fibers also do not hold their brightness as long as chemical pulps, i.e., they tend to have brightness reversion problems. North American SGW capacity is currently around 7 million metric tons per year.

Pressurized groundwood (PGW) is made similar to GW, except that it is done under pressurized conditions, allowing elevated shower water temperatures during grinding. Developed during the 1970s, the process applies pulpwood "blocks" against special-design grindstones at temperatures approaching the boiling point of water. This softens the lignin and reduces damage to the fibers during grinding. The resultant pulp is generally of much higher quality than SGW, and the process is typically not as energy intensive, requiring about a third less energy than the TMP process discussed below.

Groundwood pulps are generally not made using chips and a grindstone. However, fine groundwood pulp (FGP), is sort of an exception to this rule. In this process, invented in Japan but not widely used today, chips are ground against a highly modified grindstone at atmospheric pressures, producing the characteristic fine fibers that are processed and bleached similar to SGW.

Refiner Mechanical Pulp (RMP). This process that does use chips was invented and put into commercial practice during the 1960s. RMP fibers are generally longer and stronger than groundwood fibers, but they tend to also have lower opacity. A typical RMP process is shown in Figure 9.

In the RMP process, chips are first squeezed to remove moisture and then are fed into large refiners of various designs. These refiners, with large groove-and-bar discs spinning in opposite directions, separate the chips into individual fiber components and discharge them at atmospheric pressure.

The number of refiners used in an RMP process will vary depending on mill size and species of wood being used. The configuration shown in Figure 9 uses two primary refiners and a secondary, tertiary, and rejects refiner.

The growth of RMP allowed increased use of residual chips from a mill's process and from sawmills. Today, RMP capacity in North America is around 625,000 metric tons annually.

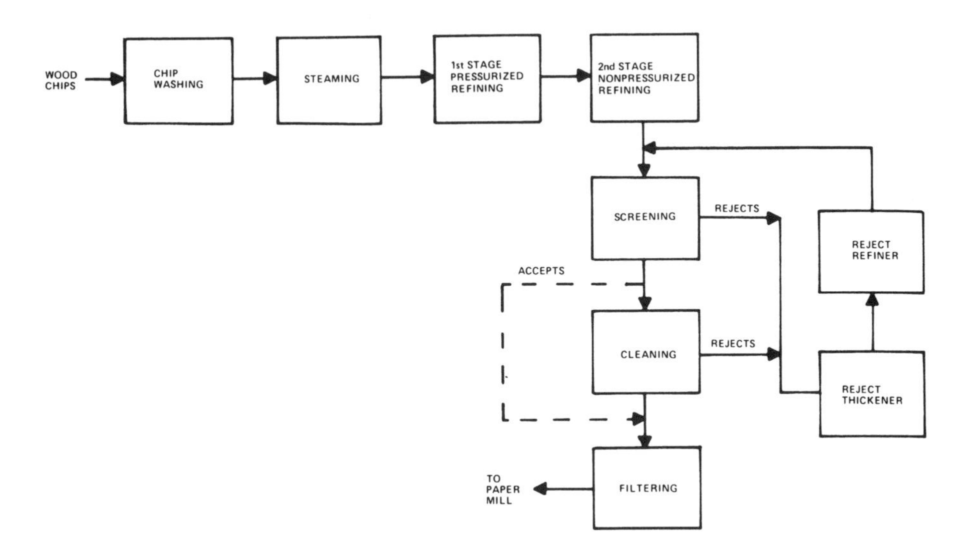


*Figure 10: Block diagram of the thermomechanical pulping process.*

THERMOMECHANICAL PULP (TMP). This process, dating back to the 1930s, was fully commercialized in the 1960s. It involves presteaming of the chips at about 230°F for several minutes, followed by pressurized refining. The chips are subjected to repeated compressions and stress relaxations in the pressurized refiner. Several stages of non-pressurized refining might followed the pressurized stage. This is followed by screening, cleaning, and washing stages. A typical TMP process is shown in Figure 10.

To make the energy-intensive TMP process more economically viable, heat generated in refining is used at high efficiency in the presteaming and refining operations. Work with this process during the 1960s advanced the technology such that the pulp being produced today is much stronger that previously thought possible with a mechanical process. Some 15 million metric tons of TMP are being produced in North America annually, including chemi-thermomechanical production.

THE CHEMI-THERMOMECHANICAL PROCESS was developed by the addition of chemicals to the chips prior to refining. Typically, sodium sulfite is added along with various chelating agents (chemical stabilizers and metal control agents) at 9–12 pH. The mixture is then preheated for several minutes at about 125°C before delivery to the refiners.

Preheating and chemical treatment tends to preserve fiber integrity in the refiner, accounting for the quality benefits of CTMP over traditional TMP. It is generally not only a longer, stronger, more flexible fiber than TMP, but also considerably brighter before bleaching. CTMP pulp yields are generally in the 90%–95% range and high ISO brightness levels in the mid- to high-80s are attainable with peroxide bleaching. CTMP provides a particularly suitable furnish for newsprint, groundwood specialties, base sheets for some of the lightweight coated publication grades, and tissue products.

The CTMP process has been further refined into related processes such as BCTMP (bleached chemi-thermomechanial pulp) mentioned above. This process uses selected northern hardwood grades, especially Aspen. All North America BTCMP capacity is in Canada, but mills in other parts of the world are also producing the grade.

Yet another variation of the TMP/BCTMP process is the APMP (alkaline peroxide mechanical pulp) system developed in the early 1990s. The APMP process involves pulping and bleaching the chips in a single stage, thus reducing equipment costs and energy requirements as much as 35%.

# Recycling and Deinking Processes

NEW RECYCLING AND DEINKING TECHNOLOGIES have been developed and implemented at a stepped-up rate during the past couple of decades. In the U.S., the government and the paper industry both set lofty recycling goals for almost every grade of paper and board. Significant gains were made during the 1990s in this regard.

Recovered papers of all types are being recycled today. OCC (old corrugated containers), from shipping boxes primarily, has been a recycling staple for many years, as has newsprint. The recovery rate for these grades are already pushing the ceiling of what can be effectively recycled from national fiber streams. Current and future challenges are in the recycling of mixed office papers. Various non-impact copy, laser, and thermal-set inks make deinking of this wastepaper source a difficult but not impossible task.

Basically, recycling of brown papers and boards that will not require bleaching is a fairly simple process. The recovered paper, typically purchased in large bales, is metered into a large pulper filled with water and various chemical agents that aid in breaking up the fibers. A "ragger" or "detrasher" apparatus extracts foreign articles such as rope, plastic, rags, cloth, etc., that invariably get through the "sorting" process when these grades are baled for recycling. The pulper has a large rotor or agitator that provides mixing and agitation.

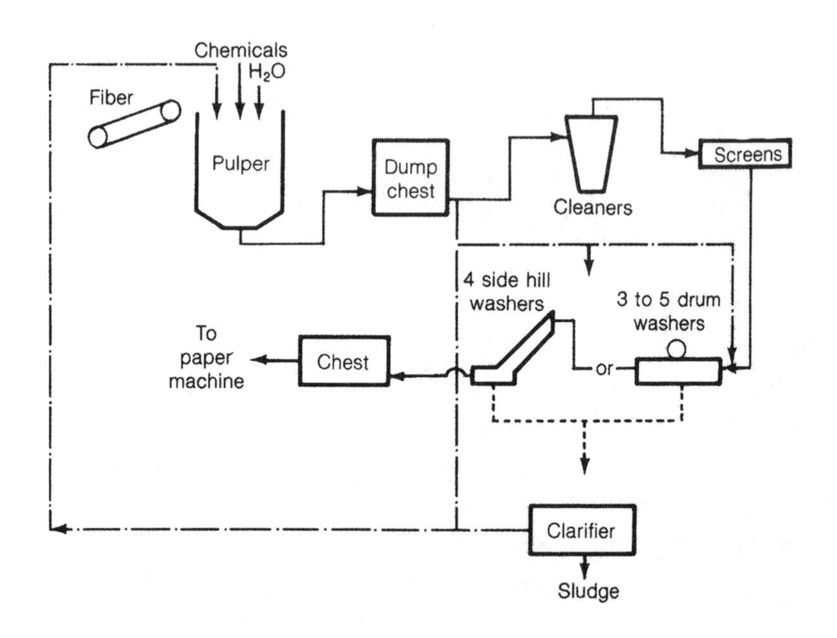


*Figure 11: Flow diagram of a typical washing deinking operation.*

After pulping, or repulping as the case might be, the brown recycling slurry goes through a series of washings, screenings, and cleanings to remove contaminants, especially "stickies," which are made up of various adhesives, plastics, tars, waxes, latex, etc. These stickies generally soften and disperse into the stock, making them very difficult to remove. As they cool, they can stick to and build up on paper and board machine components and cause holes and other defects in the finished product. Removal of stickies has been a critical technology focus in both brown and deinked grades during recent years.

In the deinking area, technologies have been generally categorized as "washing" or "flotation." Often, both approaches are employed. In the washing technique, the ink is freed from the fibers with various chemistry and washed away with large amounts of water. In the flotation process, air bubbles are injected into the wash system to lift ink particles from the fibers and carry them to the surface as a froth or foam that is removed

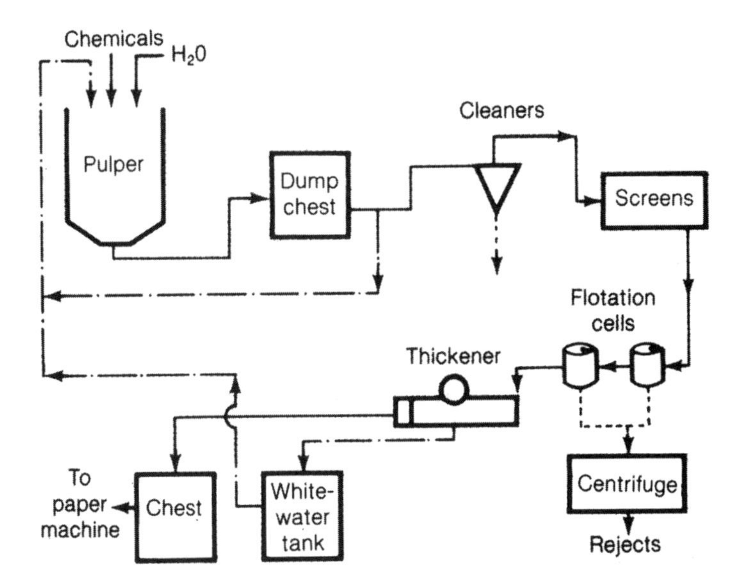


*Figure 12: Flow diagram of a typical flotation deinking operation.*

by skimming. A washing deinking arrangement is depicted in Figure 11, and Figure 12 depicts a typical flotation deinking schematic.

In the past, washing deinking was more commonly used in the U.S., while flotation deinking was more popular in Europe and other parts of the world. However, as the difficult inks mentioned above have come into wider use, the use of flotation deinking has increased dramatically in North America. Combination systems have been the design of choice during the past decade.

In the pulper of a deinked plant, common chemical additions include but are not limited to caustic soda, sodium silicate, various surfactants (detergents), hydrogen peroxide, and a chelating agent. The caustic facilitates swelling of the fibers, which acts to break away the attached ink particle. The surfactant aids in stabilizing the ink particle in solution and prevents its reattachment to the fiber.

The hydrogen peroxide pre-brightens the fiber and/or prevents it from yellowing (especially if groundwood fibers are in the mix). The sodium silicate and chelants slow the peroxide decomposition. Sodium silicate also plays other key roles in the process, such as boosting "detergency" and helping to buffer alkalinity.

In the subsequent flotation "cell," the small air bubbles attach to the ink particles that were loosened or "freed" from the fibers in the pulper. To make this process most efficient, a "vehicle" is needed to complete the attachment of bubble and ink particle. Minerals such as fine clay and calcium carbonate are typically used in this function. For that reason, mills generally include coated publication papers in the pulper feed. These grades contain high percentages of clay and carbonate in their coatings and often in the base sheets themselves. Various proprietary chemistry, including surfactant ink "collectors," are added in the flotation cell to make the ink particles more hydrophobic.

The most effective bubble size for ink particle removal is in the 30–60 micron range. Particles larger than this—generally a broad spectrum of contaminants—are usually removed by screens and cleaners further along in the process. Particles smaller than 30 microns have to be removed by washing.

Various types of washers can be used in a recycling/deinking system. One of the most common washers has been the sidehill screen, which is steeply inclined so that, as pulp slides down the large screen deck by gravity, ink is drained off with the water. Of course other types of washers are also used, including the traditional vacuum drum washer. Usually a special surfactant is added ahead of the washing stage to stabilize the ink particle in solution.

After flotation and/or washing deinking, advanced screening and cleaning are used to remove most remaining contaminants. The pulp can then be brightened, typically by oxygen and

*Pressurized peroxide bleaching tower at Great Lakes Pulp & Fibre, Menominee, MI.*

hydrogen peroxide, but chlorine containing compounds are sometimes used, depending on the fiber mix. The recycled furnish can then be sent to the paper machine.

Various types of flotation cell evolutions have entered the market in recent years. These innovations include pressurized cells, multiple-injector units, multi-chambered cells, elliptical cell designs, etc. Debates also continue over predominate approaches to deinking, generally focusing on mechanical versus chemical approaches. Today, the best approach appears to be a carefully balanced chemi-mechanical approach.

# Papermaking Technologies

THE PAPER MILL IS GENERALLY DIVIDED into separate but closely linked operations—stock preparation, the paper machine, and finishing-converting. Figure 13 shows a typical paper mill arrangement and the integration of these three functions.

The schematic in figure 13 is for a fourdrinier paper machine mentioned in "Science of Papermaking," page 13, but during the past quarter century, twinwire and multi-wire formers have become more popular, along with the newer gap formers described below.

STOCK PREPARATION. In this section of the process located just ahead of the paper machine, the pulp stock, or "furnish," as it is now called, is screened, cleaned, and conditioned for the final time before being made into paper or board. Pulp taken from high density storage in either the bleach plant or the pulp mill, or purchased market pulp (if a mill is not integrated), is sent through a series of refiners and/or other fiber "polishing" equipment known as jordans or beaters.

The stock preparation system basically shapes the fibers and removes any remaining contaminants through banks of advanced cyclone or centrifugal types of cleaners. Here the fibers are "roughened up" to improve fiber-to-fiber bonding as

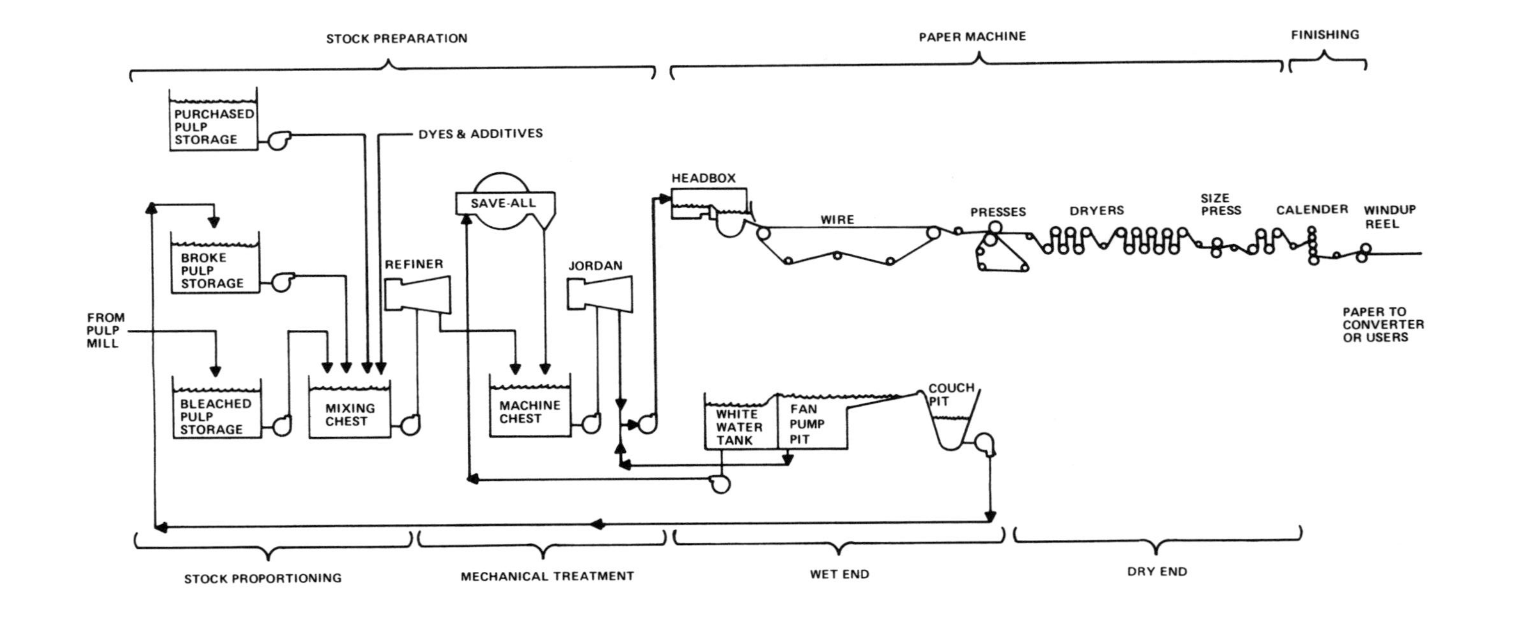


*Figure 13: Simplified flow diagram of a typical paper mill.*

*Stock preparation refiner.*

they interlock to form a sheet in the early stages of forming on the paper machine.

The strength of a sheet is generally determined by the length and interlocking characteristics of the fibers. Chemical strength additives can also be used to improve fiber bonding. In fact, many chemical agents as well as minerals and dyes are added to the furnish in stock preparation to tailor the sheet for specific customers.

Formation aids, for example, can be added to the furnish to boost the attraction of fibers and to assure a more uniform CD, or cross machine direction (width), and MD, or machine direction (length), orientation and distribution in the sheet as it is formed on the paper machine. Sizes, including organic starch, rosin, and various synthetic agents also, can be added in stock preparation to impart water resistance and desired ink printing "holdout" properties to the sheet.

Minerals are also added in stock preparation to boost opacity, brightness, and printing quality of the sheet. Additionally, these mineral "fillers," typically clay and calcium carbonates, can act to extend or replace bleached fibers in the sheet. Other pigments such as titanium dioxide, or "engineered" minerals such as calcined clay, talc, and various chemical polymers can be added during stock preparation.

The addition of carbonates also adds alkalinity and thus permanence to the sheet. This became more important in recent years as the U.S. Library of Congress and other book and document depositories around the world began discovering that some acid-based papers were literally falling apart on their shelves.

Over a long period of time, although sometimes as short as 10 years, paper made under acid conditions tends to yellow and crumble. However, alkaline papers are known to last much longer, up to 50–100 years or more. Europe got a head start in the use of alkaline mineral additions to paper beginning in the mid-twentieth century, or even earlier.

Because bleached fiber in Europe has always been more costly than in North America, due to a less abundant fiber source and higher energy costs, mills there began searching for naturally-bright fiber replacements much earlier, finding an abundance of cheap, readily assessable chalk in major "veins" throughout many parts of Europe. Permanence was not the initial driving force behind the use of alkaline minerals such as chalk, but this side benefit spurred further developments with other alkaline minerals, especially calcium carbonates.

In addition to permanence, mills discovered that alkaline mineral contents in the sheet improved general operating conditions on the paper machine and overall quality of the paper sheet. This fact pushed alkaline conversions into high gear during the 1960s, '70s, and '80s, especially in North America. Many of these conversions to alkaline papermaking were operating "nightmares" at first, but today the science is pretty well perfected.

Although most (80%–90%) of North American fine paper mills have now converted to alkaline papermaking, European mills still use higher percentages in the sheet, on the average. Today almost all European printing and writing papers are alkaline, and producers use a carbonate filler content of 18%–20% or more. U.S. mills, however, run only about 10%–12% at most, although there are exceptions.

Calcium carbonate is available in both naturally ground (GCC) and precipitated (PCC) forms. The manufacture of GCC has advanced significantly in the past decade, with ultra-fine-ground grades now being manufactured that run well in both filler and coating applications, and do not cause abrasion problems encountered with some of the earlier course-ground varieties. The GCC grades can be "engineered" for various finished paper properties.

PCC is typically produced in "satellite" plants on or near the mill property. Utilizing available mill energy, water, and carbon dioxide off-gasses, these plants operate in a "symbiotic" manner to produce precipitated carbonates with various shapes or morphologies. The main use of PCC has been as a filler, but it is finding broader applications as a key coating ingredient.

With alkaline papermaking, the traditional rosin-alum sizing combination generally cannot be used. The synthetic alkaline sizing systems most used today are ASA (alkyl succinic anhydride) or alkylketene dimers (AKD). ASA systems were initially more popular, but AKD systems are now more widely used, especially in North America, now that some shelf storage problems have been overcome.

Although most fine paper mills have converted to alkaline papermaking, some mills still continue to produce acid papers, and there will continue to be a market for these papers in the future. Typically, the filler minerals used for these papers, if any, are clays. Because it tends to be acidic, clay is usually used with rosin sizing, which is "set" or precipitated onto the fibers by highly acidic alum.

Most calcium carbonate fillers are used in what is known as "free sheet" grades, which are basically "free" of groundwood fibers, although up to 10% of free sheet content can be groundwood fiber by a generally accepted definition.

Most groundwood printing and writing grades, which include newsprint, the groundwood specialty grades, and some lightweight coated papers, use clay as a filler (if a filler is used), because groundwood furnishes tend to be acidic. However, significant progress has been made in the use of carbonates in these furnishes at more or less neutral pH. An acidic furnish will tend to dissolve carbonate and other alkaline minerals, and cause paper machine foaming problems in the process.

In addition to mineral fillers, sizing, colorants, formation aids, strength additives, etc., certain other chemical agents can be blended into the furnish to inhibit biological growth and to improve general performance on the paper machine. For example, retention and drainage aids can help the fibers drain and form faster and can help lock minerals, fiber fines, and other nonfiber components into the moving sheet, rather than washing out into the machine's "white water" loop.

Particularly as the increased use of recycled or secondary fibers has introduced more foreign contaminants and fiber fines (anionic trash) into the furnish stream, retention and drainage have become critical issues. These furnishes tend to compact more on the paper machine wire, slowing drainage in some areas and resulting in uneven CD moisture profiles.

The mineral and chemical salt elements in a furnish are collectively called "ash" because they do not combust and comprise leftover ashes when samples of pulp and paper are burned in laboratory tests to determine non-fiber contents. Keeping this ash in the sheet has become a major concern today.

*This view of the paper machine's wet end shows the headbox and fourdrinier table area.*

After the furnish has been prepared in the stock preparation area, it is ready for delivery to the paper machine. A large "fan pump" sends the furnish from the paper machine stock chest to the paper machine headbox.

PAPER MACHINE. The paper machine has three basic sections—the wet end, the press section, and the dryer section. However, it should be emphasized that the paper machine is a continuous web process and works as one integral unit. Problems in one section quickly upset those in another, and can result in "breaks," requiring time-consuming and costly (production-wise) shutdowns for rethreading and restarting of the paper machine.

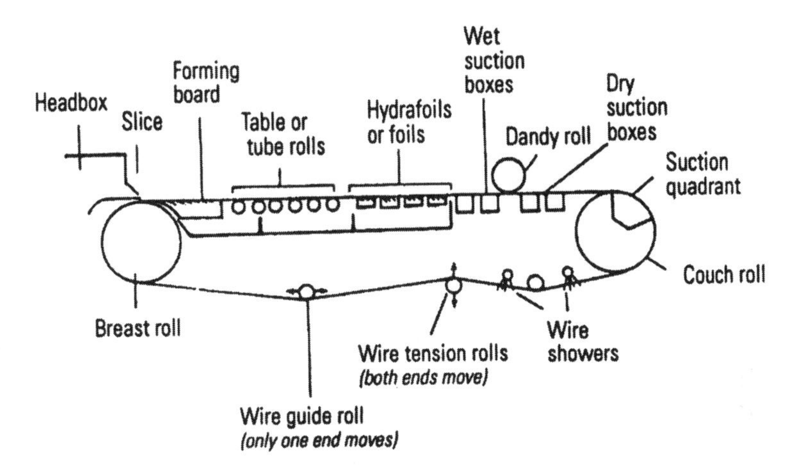


*Figure 14: Schematic of a typical fourdrinier forming section.*

WET END. The fan pump delivers mainly water to the headbox, with a 10%–20% suspension of fibers and additives. The headbox then orients the fibers in certain directions and distributes them through an opening known as a slice or lip across the width of the paper machine, which can be several hundred inches wide. This distribution must be uniform and in exact volumes relating to the type of paper (caliper or thickness) being made and specific machine speeds.

There are two basic types of headboxes in use today—the rectifier roll headbox and the hydraulic or full flow unit. The latter is more commonly used, especially on machines that have twin wire configurations. The rectifier roll headbox meters furnish onto the forming fabric by a specially designed roll, while the full flow unit uses internal chamber designs.

A recent headbox design applies furnish in a stratified or layered manner, which allows "engineering" of the sheet to some degree, i.e., certain fibers (recycled, for example) can be concentrated in the center layers, while fibers most suitable for formation (hardwood, for example) can be placed on the surfaces.

This will also allow more effective placement of minerals and other additives. These types of headboxes have been mainly applied to board and tissue grades, as a replacement for secondary headboxes. But they will likely find wider application in paper grades in the future.

The most common paper machine design for the past 100 years or so has been the fourdrinier (discussed in the "Science of Papermaking," page 13). This type of paper machine wet end, as shown in Figure 14, is a basic flat forming section with a continuous fabric loop riding on a series of table rolls or wire turning rolls. This fabric is still called a "wire," because originally these fourdrinier belts were made of woven bronze wire. Today, most fabrics are made of woven synthetic polymer filaments or stands and are more commonly called forming fabrics. Collectively, all of the fabrics (forming, pressing, and drying) are called paper machine "clothing."

The headbox deposits a layer of water-fiber to the moving wire. Water drains away through the porous fabric as it moves rapidly down the forming section, leaving a wet, weak mat of fiber and additives. The suspension first passes over the forming board, located just after a large roll under the headbox called the breast roll.

The forming board is so called because in very early fourdrinier machines it was a large wooden board that supported the heavy water-fiber suspension leaving the headbox, keeping it from damaging the wire. Today's forming boards mainly have the all-important function of setting the first formation characteristics of the sheet.

Typically, a series of forming/drainage elements called foils and vacuum boxes follow the forming board. Table rolls were once used as the primary drainage/forming element on a fourdrinier, but these have more or less been replaced by foils and vacuum boxes. A foil blade is an angled baffle with its leading edge against the wire, creating a suction effect that helps drain

*Willamette's Hawsville, KY, mill features a rare twin shoe press on a fine paper machine.*

the mat. Also, the "bump" created by this blade kicks fibers several millimeters into the air as the mat passes, and allows them to realign in the machine direction, improving surface smoothness and printing quality of the sheet.

Large vacuum pumps under the wet end pull water from the sheet as it passes over the vacuum box slits. At a certain point down the fourdrinier, the mat loses its wet, glossy look. This event is called the dry line. After the dry line, other elements can be placed above the sheet, such as a steam box or a dandy roll. The steam box applies steam to the sheet to even out moisture profiles and the dandy roll imparts certain characteristics to the sheet, including watermarking.

The final roll in a fourdrinier section is the couch roll. It sits on the other end opposite of the breast roll. The couch roll delivers the newly formed sheet into the press section.

*This supercalendar is located in Rhinelander, Wisconsin.*

Sometimes, on board grades, the couch roll "nips" against a special roll located above it, in a pre-press arrangement known as a couch press.

PRESS SECTION. Barely able to stand alone as a sheet, the fiber mat passes over the couch roll into the press section. Here the sheet passes through the nips of various sized rolls pressed against each other. The pressure on these rolls (measured in pounds per linear inch, or pli) is called the nip loading. As the sheet passes through these presses, water is progressively collected/removed by the press fabrics (also called wet felts) looping around and over the rolls.

The sheet can pass more or less in a single plane through the press section (straight-through press) or serpentine in various configurations around and through the rolls. It then exits the press section into the dryer section.

Newer "wide-nip" press section designs have focused on extending the time that the sheet is actually in the nip. This is

*The parent roll is visible in the foreground of this paper machine and the dryer section in the background.*

done in various ways, e.g., by allowing the nipped rolls to deflect (using special covers and designs) and in effect flatten out as they nip under pressure, or by special "shoe" belts that wrap the nip for several degrees. These presses, initially used on board grades but now commonly used on fine paper and other grades, can dramatically improve water removal efficiency without crushing the sheet.

The sheet leaves the press section at 35%–50% consistency, depending on the press section being used and the grade being made. It then enters the drying section for continued moisture removal.

DRYER SECTION. The typical dryer section contains several stations or sections of steam-heated drums, around which a dryer fabric carries the sheet until it is more or less "airdry." The drums are typically five or six feet in diameter, with a total of 50 or more arranged in three to five separate sections.

*The winder here operates at 8,300 feet per minute.*

*The automated line wraps 1,500 tons per day.*

*This pulp dryer includes an air-borne dryer section and a pulp sheeting system.*

The dryer fabric and sheet pass through these sections in a prolonged serpentine run. In a more recent development known as single-tier drying, the dryer drums do not have a top layer of rolls. This improves the simplicity and overall efficiency of the dryer section.

For enhanced surface properties, a size press is located between some dryer stations. Typically these stations apply starch, casein (a protein substance processed from cow's milk), and/or other sizing materials to the sheet surface. With some of the more recent supercalendered groundwood specialty grades, the size might also contain some mineral pigments.

The groundwood specialty grades are then supercalendered (tall presses with stacked polished steel rolls that "iron" the surface smooth) in an off-machine operation. The supercalendered (SC) grades that have some pigmentation and thus a higher printing quality are referred to as SC-A or SCA-plus. Some newer grades that have size press-pigment applied to the surface (usually with special "extrusion-type" film presses) and receive an on line calendering treatment with softnip calenders

*Coating in 1927.*

(calenders with special flexible surfaces) are being referred to as film coated offset papers.

After the dryer section, the paper or board generally passes through an in line calender (smaller than supercalenders) and is wound into large rolls (parent rolls) at the end of the machine. The parent roll is then conveyed to a winder/slitter unit, which slits and winds it into rolls sized for specific customers. These rolls are then wrapped and sent to the warehouse for subsequent shipment. Some mills have on site converting, where the rolls are cut into sheets, packaged, and palletized.

In the manufacture of market pulp, some mills use "airfloat"

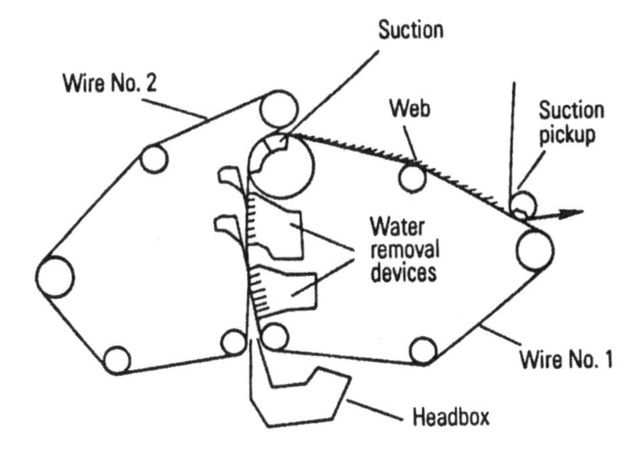


*Figure 15: A Bel-Baie twin wire forming unit.*

or "airborne" dryers, in which the sheet moves through heated drying chambers on a "bed" of hot air. In this manner, the sheet is never touched by rolls or fabrics. Other market pulp mills use standard drum-based drying sections.

MULTI-WIRE AND GAP FORMERS. As mentioned above, some fourdrinier sections, usually for board grades, can have additional headboxes located in various positions down the fourdrinier. These headboxes apply second and third plies in the formation of boards.

During the past couple of decades, the fourdrinier has been retrofitted with a topwire former, consisting of a second and separate wire-and-roll-run on top of the sheet. This reduces or eliminates the "two-sidedness" of typical fourdrinier grades, in which the side that contacts the forming fabric (wire side) is different from the side that does not (top side).

Also, separate twin wire formers have been built and installed in recent years, completely replacing the traditional fourdrinier former. A typical twin wire unit is depicted in Figure 15. These units not only eliminate wire side versus top side problems, but can operate at much higher speeds.

*This modern online machine is coating both sides of the sheet simultaneously leading into the non-contact dryer section.*

Gap formers use a twin wire arrangement in which the headbox actually injects the furnish between the two wires through an open gap of air. The forming zones on the units are typically very short, allowing the machine to operate at extremely high speeds. Twin wire units with gap formers are mainly being used in the production of high quality printing and writing papers and newsprint.

COATING TECHNOLOGIES. The coater applies a "formulation" of minerals, starch-latex binders, and other premium pigment ingredients to the finished sheet, usually giving it a

*A contemporary coating kitchen.*

glossy, highly smooth printing surface for excellent ink holdout and graphic fidelity. In the U.S., this is mainly done with on line coaters, while off line coaters are more common in Europe. However, these practices are evolving and changing as customer bases continue to shift.

Various types of coaters are in use today. With the basic blade coater, an angled, stiff blade is used to "doctor" off and smooth out excess coating materials, known as "color," which is applied to the sheet in various pond, applicator roll, or air-jet arrangements. The bent blade coater uses a bent or extended blade such that it has a troweling effect in doctoring off the color. It produces a smoother and more uniform surface than traditional stiff blade units and has less tendency to streak.

The trailing blade coater uses a flexible doctoring blade that points in the direction of the roll rotation. In a flooded nip coater, the nip between the applicator roll and its "backing roll" is filled with color to create hydraulic pressure. The fountain

*This modern integrated mill is located in Port Hawkesbury, Nova Scotia.*

blade coater uses a slot to apply or "extrude" the color just prior to the metering or doctoring blade. A rod coater uses a rotating bar to doctor excess color, turning opposite of the direction of sheet travel.

With an air knife coater, excess color applied to the sheet with an applicator roll and other methods is doctored off and smoothed by a thin jet of air. The short-dwell coater uses a pressurized pond to apply the coating color to the sheet, followed immediately by a blade. The sheet remains in actual contact with the coating color for only a very short time. Advantages of the short dwell coater include high-quality coating applications at low coat weights and fewer web breaks.

There are many other combinations of roll, blade, and extrusion coating techniques. There are also sophisticated, highly effective designs, such as the Twin Coat Tower, which has two blades, one on top of the other, to apply color in immediate succession—to the front and back of the paper with no intermediate drying.

Papers and boards can be coated on one side only (C1S) or on both sides (C2S). The color can be applied in one stage or in multiple stages, the latter being practiced more in Europe than in North America, although U.S. mills have been doing much more multiple coating in recent years, especially on board grades. Multiple coating allows the use of coarser pigments in the first layer (for opacity primarily), followed by finer premium pigments on the outside, for highest quality and economy.

The coating color is formulated and mixed in the mill's coating "kitchen." Formulations typically contain kaolin clays, ground and precipitated calcium carbonate, calcined clay, titanium dioxide, latex, starch, talc, and other pigments and polymers to maximize gloss and printing properties. GCC in particular has been increasingly used in coating formulations in recent years.

The coating applied to paper is usually dried immediately in a special drying chamber. Today, the more common dryers tend to be of the infrared variety, either electric- or gas-fired.

# Glossary

ACID: Generally, any of various typically water-soluble and sour compounds used in pulp and paper mills capable of reacting with a base to form a salt. It frequently refers to the cooking liquor used in the sulfite pulp mill.

AF&PA: See AMERICAN FOREST & PAPER ASSOCIATION.

AIR BLADE: A metal blade used in a method of coating paper that extends across the paper machine. It contains a slot which distributes air so that it removes excess coating and smoothes the surface. See also AIR DOCTOR and AIR KNIFE COATER.

AIR DOCTOR: A long, thin jet of air acting as a flexible doctor blade which meters off the controlled excess of previously applied coating on a paper coater, leaving the correct weight of coating on the sheet. See also AIR BLADE, AIR KNIFE, and AIR KNIFE COATING.

AIR KNIFE COATER: Type of coating equipment that operates on the "air knife" principle using a thin, flat jet of air for removing excess coating immediately after it is applied to the sheet. See also AIR BLADE and AIR DOCTOR.

AIR KNIFE: See AIR BLADE.

ALKALINE PAPERMAKING: The manufacture of paper on a paper machine under alkaline conditions by the use of wet end additives (fillers) such as calcium carbonate ($CaCO_3$) and neutral size. Paper made under these conditions are used particularly in special applications—documents, books, maps (where aging resistance is important), milk carton stock, cigarette papers—but operating advantages offered by the process have now prompted almost all fine paper mills (especially in Europe) to convert to the process from acid systems.

ALKALINE PROCESS: A chemical process that reduces fibrous raw materials to their individual fiber components, or the treatment of pulp and paper produced from it by the use of alkaline-based chemicals.

ALKALINE PULP: Pulp made by the cooking of chips with alkaline-based chemicals. Sulfate and soda pulps are examples.

ALKALINE PULPING: The process of cooking chips with alkaline-based chemicals.AMERICAN FOREST & PAPER ASSOCIATION (AF&PA): Headquarters: Washington, D.C., (202) 463-2700.

ALKYLKETENE DIMER (AKD): A synthetic sizing agent used for internal sizing of paper, specifically in neutral or alkaline papermaking systems.

ALUM: A papermaking chemical, $Al_2(SO_4)_3 \cdot 14H_2O$, $Al_2(SO_4)_3 \cdot 18H_2O$ or a mixture of these hydrates, commonly used for precipitating rosin size onto the pulp fibers to impart water resistant properties (when used for water treatment) to the paper made from it. Also called aluminum sulfate or papermaker's alum.

ANIONIC TRASH: See ANIONIC.

ANIONIC: (1) A negative electrical charged ionic particle. Anionic materials in liquid, when subjected to electric potential, migrate to positive pole or anode. Anionic surfactants are commonly used as polymerization emulsifiers and stabilizers in paper coatings. (2) Also refers to materials in papermaking furnish that interfere with the retention and performance of chemical additives, such as pulping and bleaching residues, recycled fiber, mechanical pulps, fillers, anti-scalants and pitch dispersants. Sometimes referred to as anionic trash.

ANTHRAQUINONE (AQ): A quinoid compound added to alkaline pulping to improve yield and to increase the delignification rate.

ASH CONTENT: The percentage of ash resulting from the complete combustion of a specific sample weight of cellulosic material, such as wood, pulp, and paper, in which all carbon, combustibles, and volatile compounds are removed. It is an indication of the amount of mineral salts and inorganic foreign matter in wood and pulp. It is also an indication of the filler, coating, pigmentation, and chemical additive content in a paper sheet.

ASH: The inorganic residue of complete combustion of paper industry cellulosic materials, such as wood, pulp, and paper, and used as a measure of some of its constituents. See also ASH CONTENT.

BAFFLE BOARD: Wooden-type dams and plates used in the headbox of a paper and board machine to prevent flocculation and eddy currents. It promotes good mixing and velocity control.

BAFFLES: Plates placed perpendicular to the flow to change direction of a flowing medium. Also, a system of dams and plates in a paper machine headbox. See BAFFLE BOARD.

BARKING DRUM: A large, horizontally inclined, rotating metal drum equipped with internal steel bars into which pulpwood logs are conveyed to remove bark by mechanical abrasion as they tumble against each other and the inner surface of the drum. Also referred to as DRUM BARKER.

BEATER: A large, longitudinally partitioned, oval tub used to mix and mechanically "work" pulp with other ingredients to make paper. It also contains a large cylindrical roll extending from one side of the partition (midfeather) and the opposite wall of the tub. The roll is fitted with steel bars (knives) on the surface under which the pulp mixture circulates between it and the bedplate below it, where the mechanical action occurs.

BISULFITE PROCESS: The process in which the cooking liquor used contains a predominance of bisulfite ion in the 2 to 6 pH range with little or no true free $SO_2$. The reduction is accomplished at elevated temperature and pressure conditions. Calcium base liquors were first used, but sodium and magnesium are more common today.

BLACK LIQUOR OXIDATION (BLO): The transformation of unstable sodium sulfide ($Na_2S$) to the more stable sodium thiosulfate ($Na_2S_2O_3$) in black liquor by the use of air or oxygen ($O_2$) so that objectionable hydrogen sulfide ($H_2S$) and organic sulfur compound formation in its evaporation and combustion are minimized, thus preventing their loss from the sulfate recovery process. The treated liquid is called oxidized black liquor.

BLACK LIQUOR RECOVERY BOILER: A boiler designed especially to recover heat by burning concentrated black liquor (from the cooking of wood by the sulfate process) and to use the heat for steam generation.

BLACK LIQUOR: Liquor from the digester to the point of its incineration in the recovery furnace of a sulfate chemical recovery process. It contains dissolved organic wood substances and residual active alkali compounds from the cook.

BLADE COATER: A paper coating machine that uses a flexible metallic blade to spread coating material on the web of paper after it has been applied by a dip roll. Also referred to as flexible blade coater or a trailing blade coater.

BLEACHED CHEMITHERMOMECHANICAL PULP (BCTMP): A high brightness mechanical pulp (generally sold as market pulp) produced by the chemithermomechanical pulping process and bleached to 90-plus G.E. Brightness. It has characteristics similar to some bleached kraft pulp grades, but with a much higher yield and generally lower capital costs for production capacity.BOILER: Broad or general term for a steam-generating unit. It is referred to as an industrial boiler when primarily used to generate steam for process requirements such as in a pulp and paper mill, or as a recovery boiler when used in the chemical recovery cycle of a pulp mill.

BLOW TANK: A storage tank in which cooked pulp is stored after it is blown tangentially from a digester into a separator, with pulp dropping into the tank, and the steam and gases escaping from the top vent to a steam condenser.

BOARD: See PAPERBOARD.

BREAK: (1) Denotes a tear completely across the sheet of paper or paperboard that occurs while the paper machine is running or during some subsequent conversion operation. (2) When referenced to an electrical circuit device in the mill, it is the minimum distance between the stationary and movable contacts when these contacts are in the open position.

BREAST ROLL: A large metal or fiberglass-covered roll located so as to support the fourdrinier wire (forming fabric) at the headbox end of a paper machine where the stock inlet admits the stock to the wire. It is usually driven by the wire which runs over it.

BROWN PAPER: Grades of paper made from unbleached pulp stock such as corrugating medium, linerboard, and kraft paper.

BROWNSTOCK: Brown-colored pulp from the cooking process and up to the bleaching process in a chemical pulp mill.

BUBBLING: A technique of introducing microbubbles into dry coating to increase its brightness and opacity.

BULK: The compactness property of a sheet in relation to its weight (whose value decreases as the compactness increases). It is measured as the thickness of a pile of a specified number of sheets under a specified pressure.

CALCIUM CARBONATE (CaCO3): (1) A white pigment commonly used in the paper industry as a paper coating material and filler. It is a compound that can be chemically prepared but also occurs in a natural form. (2) A naturally occurring mineral called limestone, usually combined with magnesium carbonate (MgCO3), and used primarily in the pulping industry to produce lime (CaO). It is used as ground calcium carbonate (GCC) and precipitated calcium carbonate (PCC) in the paper industry.

CALCIUM OXIDE (CaO): A chemical obtained by burning limestone (CaCO3), also called lime and used in the causticizing area of a pulp mill, and mixing with green liquor to convert sodium carbonate (Na2CO3) into sodium hydroxide (NaOH). The calcium carbonate precipitate (lime mud) can then be burned in a kiln to produce more lime for reuse.

CAPTIVE PULP: Pulp used by a paper mill that is made in a pulp mill on the same site and integral to it. May also refer to pulp (market pulp, in particular) made by one mill and used by another belonging to the same company. Could also apply to a "customer" who is part owner (or some other contractual arrangement) in a joint-venture market pulp mill.

CARBONATES: Types of fillers, such as calcium carbonate, that will react with acidic chemicals or will produce alkaline conditions in water solutions. Also used in coating formulations. See also ALKALINE PAPERMAKING, GROUND CALCIUM CARBONATE (GCC), and PRECIPITATED CALCIUM CARBONATE (PCC).

CASEIN: A paper sizing and adhesive coating material produced by coagulating the protein of skimmed milk with a suitable acid, such as hydrochloric, sulfuric, or lactic acid.

CAUSTIC EXTRACTION: A stage in the pulp bleaching sequence (E) that normally follows the chlorination stages to remove alkali-soluble, chlorinated lignins. It is also used in some non-chlorine bleaching sequences.

CAUSTIC SODA: Sodium hydroxide (NaOH), a chemical used as a raw material and found extensively in pulp and paper processing materials, especially in the soda and sulfate process.

CAUSTICIZING: Converting green liquor to white liquor by the use of slaked lime [Ca(OH)2] that reacts with the sodium carbonate (Na2CO3) in the green liquor to form active sodium hydroxide (NaOH) in the white liquor. Also called recausticizing.

CELLULOSE: The chief substance in the cell walls of plants used in pulp manufacturing. It is the fibrous substance that remains after the nonfibrous portions, such as lignin and some carbohydrates, are removed during the cooking and bleaching operation of a pulp mill.

CHALK: (1) A natural form of calcium carbonate (CaCO3) used to make paper filler and coating material. Sometimes called Paris white. (2) Limestone that is used to produce lime in a lime kiln for recausticizing green liquor. See CALCIUM OXIDE.

CHELATING AGENT: (1) Any chemical used to stabilize other chemical solutions, such as using magnesium sulfate to stabilize hydrogen peroxide bleaching liquor. (2) Chemicals used in the pulp bleaching process to control the brightness restricting and reversion effects of iron salts and other heavy metals in the pulp. (3) Organic chemical compounds that combine with iron and contribute to corrosion of steel digesters during alkaline pulping. Also called a sequestering agent or chelate.

CHEMICAL PULP: The mass of fibers resulting from the reduction of wood or other fibrous raw material into its component parts during the cooking phases with various chemical liquors, in such processes as sulfate, sulfite, soda, NSSC, etc.

CHEMIGROUNDWOOD: A process of producing groundwood pulp in which whole barked logs are given a pressurized heat treatment in sodium sulfite and sodium carbonate solution to soften the wood prior to grinding.

CHEMIMECHANICAL PULP (CMP): Pulp made by pretreatment of chips with chemicals at a temperature usually below 100°C, followed by atmospheric refining.

CHEMIMECHANICAL PULPING: Wood grinding or chip refining process employing a mild chemical treatment to improve strength or increase production.

CHEMITHERMOMECHANICAL PULP (CTMP): Pulp made by the thermomechanical process, in which the wood chips are pretreated with a chemical, usually sodium sulfite, either prior to or during presteaming, as an aid to subsequent mechanical processing in refiners. Sometimes referred to as thermo-chemimechanical pulp (TCMP).

CHINA CLAY: A superior type of naturally occurring clay comprised mainly of aluminosilicate mineral (Al2O2 • 2SiO2 • 2H2O) known as kaolinite or kaolin, and commonly used in conjunction with other similar materials as a paper coating ingredient or sheet filler.

CHIPPER: A piece of equipment in the woodyard/pulp mill area used to ìchipî whole logs. It consists of an enclosed, rapidly revolving disk fitted with surface-mounted knives against which the logs are dropped in an endwise direction in such a manner that they are reduced to chips, diagonally to the grain.

CHIPS: Pieces of wood (approximately 1 inch square and 1⁄8 inch thick) resulting from the cutting of pulpwood logs in chippers in the wood preparation area of a pulp mill prior to conversion into pulp in the digester(s). Sometimes called wood chips.

CHLORINE DIOXIDE (ClO2): A greenish gas pulp bleaching agent commonly used with kraft chemical pulp, but is also used to bleach other chemical and mechanical pulps, as well as some recycled furnishes. In recent years, it has been used to replace chlorine, in what is known as chlorine dioxide substitution, primarily for environmental reasons. It is typically made on site using a chlorine dioxide generator, by chemical reduction of sodium chlorate. See also CHLORINE DIOXIDE GENERATOR, CHLORINE DIOXIDE SOLUTION, and CHLORINE DIOXIDE STAGE.

CHLORINE DIOXIDE GENERATOR: A unit or system that produces ClO2 on-site in a dilute aqueous solution by the reduction of sodium chlorate in strong acid solution with a reducing agent, typically sulfur dioxide, methanol, or sodium or hydrogen chloride. Several processes are currently being used to produce $ClO_2$, which is used in the bleaching and brightening of pulps, and also in water treatment and biological control.

CHLORINE DIOXIDE SOLUTION: A very unstable water solution of chlorine dioxide gas ($ClO_2$) produced in the chemical preparation area of a pulp mill. It is used in the pulp bleaching process.

CHLORINE DIOXIDE STAGE: The step or steps in a multi-stage bleaching process ("D"-stages) where chlorine dioxide solution is mixed with pulp, allowed to react, and then washed as one of the operations making up a complete pulp bleaching system.

CHLORINE: A greenish-yellow, poisonous, gaseous chemical element ($Cl_2$) used in bleaching pulp and water purification in a pulp and paper mill.

CHLORINE: A greenish-yellow, poisonous, gaseous chemical element ($Cl_2$) used in bleaching pulp and water purification in a pulp and paper mill.

CHLOROFORM: A chlorinated organic ($CHCl_3$) sometimes produced during the bleaching of pulps with chlorine and/or chlorine containing compounds.

CHLOROTOXINS: A general reference to toxins such as dioxins, furans, and other chlorinated organics produced during the bleaching of pulp with chlorine and/or chlorine containing compounds.

CLASSIFIER: (1) A device for separating pulp fiber in liquid suspension according to fiber length and particle size by depositing the fractions on known mesh wire cloth for final drying to determine bone dry weight. (2) A device on a coal pulverizer outlet that separates oversize particles and recycles them back to the pulverizer.

CLAY: A naturally occurring, earthy, fine-grain material comprised of a group of crystalline clay minerals with a natural basic structure of aluminosilicate whose hydrous chemical form is $2H_2O \cdot Al_2O_3 \cdot 2SiO_3$. It is commonly used in the paper industry to make up paper filling and coating materials. Clays are sometimes altered by further refining, heat treatment, etc., to enhance or extend their end uses, e.g., calcined clay and delaminated clay. See also CHINA CLAY and KAOLIN.

CLOSED CYCLE MILL: A pulp mill concept in which all liquid effluents are recovered, thereby eliminating water pollution by the mill. Also called effluent-free mill, minimum impact mill, zero-discharge mill, and totally effluent free (TEF) mill.

CLUSTER RULES: U.S. EPA regulatory guidelines (promulgated November 1997) effecting the pulp and paper industry by "clustering" water, air, and solids guidelines into a single rule, although the guideline for each were actually promulgated separately and, in effect, are separate rules. The Cluster Rules are particularly restrictive in regard to methanol emissions and chlorinated organic discharges (See AOX), such as dioxins and furans.

COATER BLADE: The blade that doctors off coating color applied to paper or board, resulting in a highly uniform and smooth surface of a specific thickness or coated weight.

COATING KITCHEN: An area or room in a paper mill located in the stock preparation area where paper coating and coloring materials are stored, prepared, and mixed in proper proportions. Sometimes referred to as color kitchen, color room, or mixing room.

COATING: (1) The process of treating a sheet of paper or paperboard so that a pigmented coating material layer or a clear film is applied to its surface. (2) Refers to the coating material or film substance before coating. (3) The coating layer that is formed on the paper and paperboard sheet.

COGENERATION: Generation of power in an industrial power plant to produce both steam and electricity for in-plant use, as well as for sale to outside utility companies.

CONICAL REFINER: A type of mechanical pulp refiner consisting of a conical plug whose surface is equipped with individual bar and wood spacers, and is fitted inside a similarly equipped shell between which the pulp slurry is moved as the motor-driven plug rotates. It applies a refining action on the fibers and is commonly known as the jordan refiner.

CONTINUOUS DIGESTER: A wood-cooking vessel in which chips are reduced to their fiber components in suitable chemicals under controlled temperature and pressure in a continuous operation, as compared with a non-continuous batch digester.

CONTINUOUS: (1) Pulp and papermaking processes and process equipment that operate to completion in an uninterrupted series of actions or events. (2) An entire sheet of paper in a roll.

CROSS DIRECTION (CD): A term referring to the side-to-side direction of a paper machine, or to the paper sheet made on it, as opposed to along the machine or machine direction which runs from head to exit end. Sometimes expressed as across machine direction. See also CROSS GRAIN.

CROSS MACHINE DIRECTION: See CROSS DIRECTION.

CYCLONE CLEANER: A pulp stock cleaner in which pressurized slurry is tangentially injected into a truncated, cone-shaped vessel to produce a rapid spinning action of the suspension, causing a centrifugal force that moves the heavier particles to the outside wall where they drop down to the bottom and are removed. The inner vortex column with the acceptable fiber moves to the top where it is removed for further processing. Also called a vortex cleaner.

CYLINDER MACHINE: A type of papermaking machine primarily used to make paperboards. The forming cylinders are constructed or covered with wire so that, as they turn within a tank or vat particularly filled with a stock solution, fibers are picked up to form a web on the surface, with water draining through and passing out at the ends. The wet sheet is then transferred off the cylinder onto a felt for possible combining with the other sheets (multiple cylinders on same machine) and subsequent pressing and drying. Sometimes called a vat machine.

D100: Total (100%) substitution of chlorine with chlorine dioxide in a pulp bleach plant. The sequence or plant itself is said to be ìD100.î

DANDY ROLL: A woven wire cloth skeleton cylinder usually located on top of the paper machine wire (forming fabric) and just ahead of the couch roll. Sometimes covered with an arrangement of thin longitudinal wires, crossed at close intervals by heavier, circumferentially wound wires used to produce a wave and laid effect on the paper surface. Designed with numbers and figures in the wires to impart watermarks on the sheet while wet. See WATERMARK DANDY ROLL.

DEFIBRATED PULP: Pulp made by mechanically reducing chips into their fiber components at elevated temperatures, usually after presteaming in a preceding chamber.

DEINKING: The processing of printed and other used/reclaimed wastepaper by mechanical disintegration, chemical treatment, washing, and bleaching to remove ink and other undesirable materials so that it can be reused as a source of papermaking fiber.

DELIGNIFICATION: Separation of the lignin components from the cellulose and carbohydrate materials of wood and woody materials by chemical treatment and washing, such as the cooking of chips in a digester (usually elevated temperature and pressure steam conditions) and subsequent bleaching of the cooked pulp. See also OXYGEN DELIGNIFICATION.

DIGESTER: (1) A pressure vessel used to chemically treat chips and other cellulosic fibrous materials such as straw, bagasse, rags, etc., under elevated temperature and pressure in order to separate fibers from each other. It produces pulp. (2) In a waste treatment plant, it is a closed tank that decreases the volume of solids and stabilizes raw sludge by bacterial action.

DIRECT CONTACT EVAPORATOR: Type of black liquor evaporator

used just ahead of the recovery furnace that concentrates the liquor by removing water through direct contact with the hot furnace flue gases.

DRAINAGE AIDS: Substances added to pulp stock slurry to increase its capability to lose water by gravity when on a screen or paper machine wet end wire (forming fabric).

DRY LINE: The line at which the free, glossy, water surface disappears on the forming wire (fabric) of a fourdrinier paper machine.

DRYER GROUP: See DRYER SECTION.

DRYER SECTION: A grouping of a number of gear-driven, cylindrical metal drums in series that are steam-heated to dry wet paper passed over it. The dry end of a paper machine may contain several of these sections. Sometimes referred to as DRYER GROUP.

ECF PULP: Pulp that has been bleached with an elemental chlorine free bleaching process. See ELEMENTAL CHLORINE FREE.

ECF: See ELEMENTAL CHLORINE FREE.

EFFLUENT CLOSURE: See CLOSED CYCLE MILL.

EFFLUENT: Pulp or paper mill discharge that tends to pollute the environment, particularly water streams, lakes, and other bodies of water.

ELEMENTAL CHLORINE FREE: A reference to pulp bleaching sequences that do not use elemental chlorine in any stage (sequences that employ no chlorine containing compounds at all are known as totally chlorine free). Most commonly, ECF pulp is a kraft chemical pulp that has been bleached to market brightness with chlorine dioxide as the primary chlorination agent, rather that elemental chlorine. See also D-100, NON-CHLORINE BLEACHING, SUBSTITUTION, and TOTALLY CHLORINE FREE.

EVAPORATORS: Process equipment in which various liquids, such as spent liquors from the pulp washing process, are concentrated by the heat removal of water so that chemicals may be readily removed from it.

EXTENDED DELIGNIFICATION: The carrying out of the delignification process further than what is normally done to produce wood pulp in order to obtain stock intended for special purposes, or to improve its subsequent bleachability.

EXTENDED MODIFIED CONTINUOUS COOKING (EMCC): A method of extending the pulp delignification capacity of a Kamyr continuous digester by applying white liquor in the final high temperature wash stage, comprising, in effect, a third cooking zone in the digester. See MODIFIED CONTINUOUS COOKING (MCC).

EXTRUSION COATING: Coating of paper or paperboard with heated, solvent-based plastic materials.

FAN PUMP: A large centrifugal-type pump used to pump and mix pulp stock and large quantities of recirculated dilution water and stock to the headbox of a paper machine.

FIBER: An elongated, tapering, thick-walled cellular unit that is the structural component of woody plants. Fibers are separated from each other during the pulping operation and reassembled into the form of a sheet during the papermaking process in the paper mill. Frequently spelled "fibre."

FIELD BUS: A communications wiring apparatus or approach that allows multi-parameter and multi-drop possibilities (multiple devices on a single run of wiring). Implementation of field bus requires considerable standardization of control appliances and technologies, and thus field bus progress has been slow in the pulp and paper industry in recent years. The technology offers significantly reduced overall costs for process control devices. Currently there are two types of communication busses: one that operates at 31.25 khz (H1) and another that operates at 1 to 2.5 mhz H2).

FILTRATE: Pulp washer and filter effluents.

FINISHING: The processing of paper after the papermaking operations are completed. Included are supercalendering, slitting, rewinding, trimming, sorting, counting, and packaging prior to shipment from the paper mill.

FLOTATION DEINKING: The process of removing ink from used paper by aerating chemically treated pulped paper with vigorous agitation to cause the separated ink particles to adhere to the air bubbles that are formed and rise to the surface as froth.

FLOTATION: (1) A method of raising suspended matter to the surface of a liquid by means of various procedures, including aeration, vacuum, the evolution of gas, chemicals, electrolysis, heat, or bacterial decomposition. (2) The process of fiber entanglement and aggregate formation with a tendency to rise or float in low consistency pulp stock slurries.

FORMATION AIDS: Materials added to stock furnishes going to the paper machine that improve the attraction of fibers and assure more uniform orientation and distribution in the sheet.

FORMER: Types of machines used primarily in the manufacture of multiply paperboards. They are adaptations of and incorporate design features of fourdrinier- and cylinder-type paper machines in that a cylindrical forming roll is used to support the forming wire mesh and stock is admitted to the wire surface through nozzles and special-shaped inlets. It is also a general term referring to a variety of specialized wet end units that apply stock and form sheets for many grades of paper and paperboard.

FORMING BOARD: A plastic-covered, sometimes perforated board located under the paper machine wire (forming fabric) between the breast roll and first table roll. It is used to eliminate the sag caused by impact of the stock jet from the slice and to improve flow onto the wire.

FORMING FABRIC: Continuous screen belt made of plastic fibers and used on paper machine wet ends where the paper sheet is formed from a pulp stock slurry. Also called forming wires, plastic wires, or simply wire. At one time metal "wires" were commonly used (and some are still in use).

FOURDRINIER MACHINE: The papermaking machine invented by Louis Robert, but named after Henry and Sealy Fourdrinier who financed its development. It is used interchangeably to refer to only the wet end section onto which the pulp stock is fed and formed as well as to the entire paper machine, including the press section, the dryer section, and the calender section. See also FOURDRINIER WIRE.

FOURDRINIER WIRE: A continuously traveling, endless, woven, metallic, or plastic screen belt located in the wet end section of a fourdrinier-type paper machine. The pulp stock is fed onto the wire so that the water can be drained from it as the fibers become oriented to form a continuous web. Also known simply as a wire. See also FORMING FABRIC.

FREE SHEET: A sheet of paper that contains no mechanical pulp fibers or is made of pulp that has been subjected to the very minimum of refining or hydration, allowing the water to drain quickly when the sheet is formed on the fourdrinier wire (forming fabric). In actuality, some "free" sheets may contain small amounts of mechanical pulp. See GROUNDWOOD FREE. Also known as woodfree.

FURNISH: The materials in a pulp stock mixture such as the various pulps, dyes, additives, and other chemicals blended together in the stock preparation area of the paper mill and fed to the wet end of the paper machine to make the paper or paperboard. Commonly called stock.

G.E. BRIGHTNESS: A U.S. measure of paper or pulp whiteness or brightness using a General Electric test instrument. It is the reflectance value relative to a standard (magnesium oxide) in the blue region of visible light, at a specific wavelength of 457 nm, expressed as a percentage of the standard's brightness. The test is described in TAPPI standard T452. It is conducted by illuminating a specimen at 45% and observing it normal to the plane of the specimen over a circular area about 12 mm in diameter (thus being sometimes referred to as "directional brightness"). See also ELREPHO BRIGHTNESS, which is the brightness measure used in most other parts of the world. There is no direct relationship between the two brightness measures, although Elrepho brightness tends to be 0.5% to 1.0% higher on the average.

GAP FORMER: A type of twin wire former in which the headbox slice jet impinges directly into the gap of the unit's two converging wires (forming fabrics). There is little or no traditional forming zone in these types of arrangements.

GREEN FIELD MILL: A new pulp and/or paper mill built on a site on which no other mill exists. Also called a grass-roots mill.

GREEN LIQUOR: A liquid that is formed during the sulfate chemical recovery process by dissolving smelt from the recovery furnace in a dissolving tank. The clear liquid takes on a greenish tinge.

GROUND CALCIUM CARBONATE (GCC): Minerals mechanically refined from limestone and marble and used as a filler pigment, paper coating ingredient, etc. Sometimes referred to as natural ground calcium carbonate (NGCC). See also CALCIUM CARBONATE.

GROUNDWOOD (GW): Pulp and paper made up of mechanical fibers produced by the grinding of pulpwood.

GROUNDWOOD FREE: Any type of paper that contains less groundwood than the paper industry standard of 5 percent, determined by microscopic analysis of a properly dye-stained sample. See also FREE SHEET.

GUN: A burner or device in a furnace used to introduce oil or gas fuel (and concentrated black liquor) and air with the proper velocity, turbulence, and concentration to produce proper ignition and combustion of the fuel within the furnace.

HEADBOX: A flow chamber located at the head end of a fourdrinier or cylinder-type paper machine. It receives the diluted pulp stock slurry and regulates the head or level to provide uniform and even flow across the width of the wire (forming section) and the correct volume of stock to the screens and mixing boxes before the cylinder machine vats. It is also known as a mixbox on cylinder-type machines.

HEMICELLULOSE: The alkali-soluble, noncellulosic polysaccharide portion of the wood cell wall.

HYDRAULIC: A general reference to mill equipment operated, moved, etc., by means of a liquid force.

HYDROGEN PEROXIDE (H2O2): An oxidizing type of bleaching chemical commonly used in bleaching and deinking paper stock as well as in bleaching other chemical and mechanical pulp.

HYDROPHOBIC: The physical property of substances that make them strongly repellent of water and said to be "not" easily wetted. It is the opposite of hydrophilic.

INFRARED DRYING: Use of dryers that emit infrared rays for paper drying purposes, particularly for drying after a coater station. The infrared heating can be either electrical or gas fired, generally utilizing special ceramic elements.

IMPREGNATION: (1) The process of treating fibrous materials with penetrating liquids prior to chemical or mechanical pulping. (2) The application of liquids or molten materials to a paperboard sheet so that it will be absorbed to impart certain desired characteristics.

ISO 9000: A reference to the 9000 series of quality standards-as maintained by the International Standards Organization. Basically, these standards–adopted by national standards agencies and organizations in more than 50 countries–focus on manufacturing, installation, servicing, etc., rather than the finished products directly. The objective is to assure consistency at every stage of manufacturing and implementation. Generally, ISO 9001 certification applies to design, development, installation, and servicing stages. ISO 9002 covers production and installation, and ISO 9003 certifies final inspection and testing procedures only.

JORDAN REFINER: A type of fiber processing machine consisting of an adjustable conical plug, surface-lined with metallic bar knives, and fitted into a shell with the inner surface having similar knives. It is used to mechanically work pulp stock pumped between them as the plug is rotated in order to impart the desired physical properties to the stock and the sheet produced from it. It is a member of the conical refiner group. See also CONICAL REFINER.

KAOLIN: A naturally occurring white form of anhydrous aluminum silicate clay mineral that is refined and added to paper stock furnish, increasing the opacity and other characteristics of the sheet. It is also used to make a paper coating mixture. See CHINA CLAY.

KAPPA NUMBER: A value obtained by a laboratory test procedure for indirectly indicating the lignin content, relative hardness, or bleachability of higher lignin content pulps, usually with yields of 70 percent or more. It is determined by the number of milliliters of tenth normal permanganate solution (0.1 $KMnO4$) absorbed by 1 gram of oven dry pulp under specified conditions, and is then corrected to 50 percent consumption of permanganate. See also K NUMBER.

KNOTS: (1) A hard lump in pulpwood logs where a branch grew out of the tree. They are difficult to cook in the manufacture of chemical pulp. (2) Lumps in pulp stock caused by fibers that have not been completely separated, or specks in rag content paper caused by knotted threads.

KRAFT COOKING LIQUOR: A chemical mixture consisting primarily of sodium hydroxide (NaOH) and sodium sulfide ($Na_2S$). It is used to cook wood chips and convert them into wood pulp. Sometimes called sulfate cooking liquor. See also ALKALINE COOKING LIQUOR.

KRAFT PROCESS: The sulfate chemical pulping process. Also any equipment used as well as any intermediate or final products derived from the process. It means "strength" in German, and is a common pulp mill name for the sulfate process.

KRAFT RECOVERY CYCLE: The series of unit processes in a sulfate pulp mill in which the spent cooking liquor is separated from the pulp by washing, concentrated by evaporation, supplemented to make up for lost chemicals, and burned to recover other chemicals. These recovered chemicals are converted to new cooking liquor by reacting them with fresh and recovered lime in a causticizing operation.

LIGNIN: A brown-colored organic substance that acts as an interfiber bond in woody materials. It is chemically separated during the cooking process to release the cellulose fibers to form pulp, and is removed along with other organic materials in the spent cooking liquor during subsequent washing and bleaching stages.

LIME MUD: The primarily calcium carbonate ($CaCO_3$) sludge that settles out and is separated from the white liquor during the clarification operation in the causticizing process in a pulp mill recovery cycle prior to pumping over to the lime recovery area. Also called white mud.

LIME SLAKING: The mixing of lime (CaO) with water ($H_2O$) to form lime water or calcium hydroxide [$Ca(OH)_2$]. In the causticizing process of a sulfate pulp mill recovery cycle, it is the mixing of lime with green liquor.

LIQUOR SPRAY GUNS: Liquor-spraying nozzles through which concentrated pulp mill liquor is sprayed into a recovery furnace to be burned. See also GUN.

LIVE BOTTOM: The bottom of a storage chest or bin that is designed so that mechanical motion, usually supplied by a screw or drag chain conveyor, can be imparted to aid in removal of stored materials.

LONGWOOD LOGS: Sections of trees prior to being cut up into pulpwood lengths. Same as long log.

MACHINE DIRECTION (MD): A term referring to the direction from the wet end to the dry end of a paper machine or to a paper sheet parallel to its forward movement on a paper machine. It is opposed to cross direction, which runs from side-to-side of the paper machine or the sheet made on it. Also commonly referred to as along machine direction, grain direction, or long direction.

MECHANICAL PULP: Pulp produced by reducing pulpwood logs and chips into their fiber components by the use of mechanical energy, via grinding stones, refiners, etc.

MECHANICAL PULPING: The converting of pulpwood logs and chips into pulp by the use of mechanical energy. See MECHANICAL PULP and MECHANICAL TREATMENT.

MECHANICAL TREATMENT: (1) The development of certain desirable characteristics in pulp slurries by subjecting them to mechanical energy in equipment such as beaters and refiners. (2) The reduction of wood chips into their fiber components by subjecting them to mechanical energy in equipment such as beaters and refiners.

METHYL MERCAPTANS ($CH_3SH$): The reduced sulfur compound formed during the cooking of wood in a sulfate pulp mill and released in mill gaseous effluents. It is primarily responsible for the characteristic offensive odors associated with the sulfate mill environment.

MICRO KAPPA NUMBER: The result of a bleachability laboratory test on wood pulp to determine the bleachability or the amount of lignin by a modification of the standard kappa number test. It is performed so that it can be used on very small samples and semibleached pulp with low permanganate consumption.

MODIFIED CONTINUOUS COOKING (MCC): A modified continuous digester system (Ahlstrom/Kamyr) in which chips cooked in the vessel's traditional upper portion (cocurrent cooking) pass through a second (countercurrent) cooking zone. In some extended modified continuous cooking (EMCC) systems, the chips are cooked in yet a third zone—the traditional "high-heat" wash zone where white liquor is added and upflows through the chip mass, resulting in the EMCC feature. Both modifications improve delignification efficiency. The "gentler" cooking resulting from extended times in the digester and multiple applications of cooking liquors (rather than in one harsh "shock" stage) reportedly produces pulps with higher strength and considerably lower kappa numbers than pulps cooked in traditional, single-cooking-stage continuous digesters.

MULTIPLE COATING: The application of a paper or board coating in more than one stage, using multiple coater heads.

MULTIPLE EFFECT EVAPORATOR: An evaporator system consisting of a series of individual evaporator bodies arranged so that the vapor generated from one evaporator body becomes the steam supply to the next evaporator in the series. It is commonly used to concentrate diluted spent cooking liquor.

MULTI-WIRE MACHINE: A paper machine having more than one forming wire (forming fabric) in various arrangements, such as a twin-wire former, topwire former, multi-fourdrinier, etc.

NEUTRAL SULFITE SEMICHEMICAL (NSSC): A chemical wood-pulping process using a neutral sodium sulfite-sodium carbonate cooking liquor. The cook is carried out under slightly alkaline conditions, but with the final separation of the fibers made by unpressurized means after the initial cook has been completed. Also referred to simply as semichemical pulping process.

NEUTRAL SULFITE: A chemical wood pulping procedure in which the cooking liquor is made up of sodium sulfite, with enough sodium carbonate to maintain a pH of 7 or slightly above during the cook. See MONOSULFITE PROCESS or NEUTRAL SULFITE SEMICHEMICAL (NSSC).

NEWSPRINT: A grade of paper, combining high percentages of groundwood and/or mechanical pulps, made especially for use in the printing of newspapers.

NIP: The contact area between two rolls on pulp and papermaking machines, such as wet presses, coaters, calenders, supercalenders, etc.

NK FLOTATION TEST: A common paper mill laboratory test used to determine the size resistance of paper. It consists of floating a 2-inch square sample on standard writing ink and timing the period when the color shows through the upper surface.

NON-CHLORINE BLEACHING: The process of bleaching pulps to "market" brightness without the use of elemental chlorine or chlorine containing compounds. Such sequences typically employ oxygen, ozone, hydrogen peroxide, and other peroxygens.

NONWOOD FIBERS: Pulp and papermaking fibers obtained from plants, such as straws, bamboo, sugarcane (bagasse), reeds, rice, cotton, etc., and not from pulpwood trees.

OFF LINE: Not done in tandem with a manufacturing line, as opposed to being on line or in-line with the process. Also refers to process measurements made away from and not on line and in real time with the process.

OFF-MACHINE COATING: The process of applying coating material to a sheet of paper or paperboard in a location that is away from the machine on which it is made. Also called conversion coating.

OLD CORRUGATED CONTAINERS (OCC): A valuable source of fiber for recycled combination paperboards. OCC includes container plant cuttings as well as used corrugated containers.

OLD MAGAZINE GRADES (OMG): Used and undistributed magazines

that are a key wastepaper or recovered paper supply for the pulp and paper industry. They contain high quality fibers and mineral filler and coating that are useful in flotation deinking processes.

OLD NEWSPAPERS (ONP): Newsprint wastepaper used as a source of secondary fiber for recycling paper mills. ONP includes old newspapers collected either from households or as over-issues at newsstands or in-plant. Also includes groundwood paper trim.

OLD PAPER: Wastepaper used as a source of fiber for the manufacture of some paperboard and chipboard. See WASTEPAPER.

ON LINE: (1) The type of operation of a computer and other instruments that is actively monitoring and controlling a process or operation. (2) A manufacturing component that is placed in-line with the manufacturing process (on-machine) as opposed to being off line or off-machine. (3) A reference to having been put into operation, such as a piece of equipment that has come on line.

OPACITY: That ability of substances such as paper, flue gases (smoke), and liquids to resist the transmission of both diffuse and non diffuse light through it. It prevents show through of dark printing in contact with the backside of a sheet of paper.

OXYGEN DELIGNIFICATION: The use of oxygen to further break down the lignin bond between fibers in wood. It generally is applied in a separate stage following chemical pulping (cooking) in a digester system. The oxygen is applied in a reactor vessel, followed by extensive washing. Kappa number of cooked chemical pulps can be reduced 10 to 15 points by this treatment with oxygen.

OZONE (O3): An allotropic form of oxygen produced by passing air or oxygen through an electrical discharge (ozone generator). It is being used in the bleaching of pulp to very high brightness, representing a significant environmental gain over the use of chlorine and chlorine containing compounds in the bleach plant (resulting in the reduction of chlorotoxins). Very little or no washing is generally required after an ozone stage, cutting down on the volume of effluent that has to be treated or recovered. Since chlorine is potentially eliminated from the bleaching process (and thus a reduction or elimination of sodium chloride buildup), the use of ozone as the primary bleaching agent may help open the door to a closed cycle mill. See also NON-CHLORINE BLEACHING and OZONE STAGE.

OZONE STAGE: The application of ozone as a separate stage in the bleaching of pulp. See OZONE.

PAPER MACHINE: The primary machine in a paper mill on which slurries containing fibers and other constituents are formed into a sheet by the drainage of water, pressing, drying, winding into rolls, and sometimes coating.

PAPERBOARD: A thick, heavy-weight, rigid, single, or multi-ply type of paper traditionally made on multi-cylinder paper machines, but is now also made on fourdrinier-type machines, with and without dual headboxes or multi-former arrangements. Thickness and material vary, depending on its end use. It is used for wrappings, packagings, boxes, cartons, containers, advertising and merchandising displays, building construction, etc. Also known simply as board.

PAPYRUS: A plant used by ancient Egyptians and others for writing and record-keeping purposes. It was made up into layers and matted into sheets. The term paper was eventually derived from the name of this plant.

PARCHMENT: Writing sheet material made from the skins of animals, such as goats and sheep.

PEROXYGEN: A general reference to various oxygen-based compounds used in the non-chlorine based bleaching of pulp, such as hydrogen peroxide, peracetic acid, etc.

PLANTATION: Woodlands that are planted and managed for maximum production.

POLYSULFIDE (PS) PULPING: A sulfide pulping process in which dissolved sulfur in the white liquor is used for sulfur makeup.

PRECIPITATED CALCIUM CARBONATE (PCC): A wet end filler used in alkaline papermaking and also in some coating formulations. It generally has good opacity and bulking characteristics, and can be produced in several particle morphologies and sizes. Most PCC being used in the paper industry is made in on-site satellite plants that utilize available mill carbon dioxide streams as well as mill power and water resources. See also CALCIUM CARBONATE.

PRESS SECTION: That part of a paper machine located between the wet end section and the dryer section where water is removed by passing the wet web between rolls and felts while applying a combination of pressure and vacuum. See also PRESS.

PRESS: A paper mill term for a pair of rolls, usually located between the paper machine wire (forming) section and the dryers, through which the newly formed paper sheet is passed. It is usually installed in three sets and used for pressing out water and smoothing the newly formed paper sheet. It is also used for the application of surface coatings on the paper machine. See also PRESS SECTION.

PRESSURIZED GROUNDWOOD (PGW): Mechanical pulp made on a stone grinder where the whole grinder casing is pressurized, and increased shower water temperature is used.

PRESSURIZED REFINING: Mechanically treating pulp in a closed refining system in which the refiner operates under positive pressure. See also PRESSURIZED REFINER MECHANICAL PULP (PRMP).

PULP: A fibrous material produced by mechanically or chemically reducing woody plants into their component parts from which pulp, paper, and paperboard sheets are formed after proper slushing and treatment, or used for dissolving purposes (dissolving pulp or chemical cellulose) to make rayon, plastics, and other synthetic products. Sometimes called wood pulp.

PULPER: A disintegrator made up of a vessel fitted with a suitable agitator and used in a paper mill to break up and defiber dry intramill broke, purchased pulp, wastepaper, etc., in water to form a slurry. This slurry is further treated mechanically and chemically (to improve its quality) prior to converting to paper and paperboard. Also called a repulper.

RAG CONTENT PAPER: Paper containing from 25 to 75% cotton or rag fibers; usually includes bond, ledger, and specialty papers.

RAPID-DISPLACEMENT HEATING (RDH): A modified kraft batch-pulping method (Beloit Fiber Systems Div.) that involves the displacement of spent cooking liquor and recovered heat used in a cook so that they can be used in a later cook. Initially, the objective of RDH was to reduce the amount of heat used, to save energy and reuse chemicals. But more recently, bleached pulp mills have employed the method to extend delignification in the pulp cooking stage and reduce subsequent bleach chemical demand. The method is similar to SuperBatch (Sunds Defibrator), with which it shares common developmental origins.

RAW MATERIALS: Any materials, such as wood, water, chemicals, dyes, additives, fuels, etc., brought into a pulp and paper mill that are required and used in the production of pulp and paper or used in producing products for use in pulp and paper manufacturing.

RECOVERED PAPER: A term that is the same as and is replacing wastepaper. See also OLD PAPER, RECYCLED PAPER, and WASTEPAPER.

RECOVERY BOILER: A combination boiler unit in a pulp mill used to recover chemicals from the spent cooking liquor and to produce steam.

RECTIFIER ROLL: A roll located just ahead of the slice in some paper machine headboxes that is used for stabilization and conditioning of the stock flow.

RECYCLED FIBER: That component of a paper or paperboard furnish that is derived from recovered paper, recycled paper, or wastepaper.

RECYCLED PAPER: Any used paper returned to the paper mill as a source of fiber. It is usually mixed with other unused or virgin fibers to make up various grades of paper. The term is sometimes used interchangeably with recovered paper, wastepaper or old paper. When reprocessed into pulp, recycled paper is known as recycled fiber.

RECYCLING: The return of processed or used materials, such as fiber, paper, water, and some chemicals, back to the original process to make a new product.

RED LIQUOR: Spent cooking liquor (condensed pink liquor) that is separated from the pulp after cooking with magnesium-base liquor (magnesium bisulfite), and contains the dissolved constituents of the wood. It is subsequently sent to an incineration process to recover chemicals for reuse in making up fresh cooking liquor.

REFINER MECHANICAL PULP (RMP): Pulp made by processing untreated wood chips in mechanical atmospheric refiners.

REJECTS REFINER: A type of refiner used to mechanically treat and refine rejected fiber from stock screening and cleaning operations in a pulp and paper mill.

RETENTION AIDS: Materials such as vegetable gum, cationic starch, potato starch, sodium aluminate, colloidal animal glue, acrylamide resin, etc., added to the papermaking process at the paper machine headbox, fan pump, or other location close to the wire. They are added in small amounts for the express purpose of maximizing the retention of fillers by altering their electrical charge or bonding.

RETENTION: The retaining of dyes, additives, fillers, fiber fines, and other materials added to the stock slurry furnish in the sheet formed from it. It is expressed as a percentage of the total amount originally added.

ROSIN: A material made up of a suspension and used for internal sizing of paper and paperboard. It is obtained as a residue from the distillation of gum from resinous southern pines. Sometimes called colophony.

ROUNDWOOD: Logs as delivered to a pulp mill, with bark attached and cut to specified lengths, usually up to 10 feet.

SATELLITE CHIP MILL: Mills remotely located from pulp mills at the source of wood (the woodlands) which process chips from treelength stems. The chips are subsequently transported to the pulp mill site.

SC PAPER: Paper with supercalendered surfaces. Sometimes written as "S/C paper". It usually refers to a groundwood specialty grade of paper. See also SC-A PAPER, SCA-PLUS PAPER, and SC-B PAPER.

SC-A PAPER: A value-added groundwood specialty grade of SC paper that generally has been surface sized with a metering size press that may also include some light pigmentation in the sizing material, followed by off-line supercalendering. An SC-A paper may also contain mineral filler. See also SC PAPER, SCA-PLUS PAPER, and SC-B PAPER.

SCA-PLUS PAPER: A value-added groundwood specialty grade of SC paper that has been surface sized with starch and/or other sizing agents (containing clay and/or carbonate pigments) using an on-line metering size press and off-line supercalendering. The sheet is typically highly filled, mainly with clay (and sometime carbonate) and is produced in the 32-50 lb basis weight range, with a G.E. brightness form the mid-60s to the mid-70s, and gloss values typically exceeding 50%. See also SC PAPER, SC-A PAPER, and SC-B PAPER.

SC-B PAPER: A groundwood specialty grade of SC paper that typically receives less surface sizing and is generally of somewhat lower quality that SC-A paper. See also SC PAPER, SC-A PAPER, and SCA-PLUS PAPER.

SCREW CONVEYOR: A short-distance material mover used in a pulp and paper mill. Its primary moving element is a rotating spiral that pushes solid material, such as pulp, chips, sawdust, coal, etc., from one point to another. Also called spiral conveyor.

SEMICHEMICAL PULPING PROCESS: A two-step pulping process that uses a mild liquor, such as neutral sodium sulfite/sodium carbonate solution, for partial softening of chips, followed by a final separation of fibers by mechanical action. In this case, it would be referred to as a neutral sulfite semichemical (NSSC) pulping process.

SHIVES: Uncooked or undisintegrated bundles of fibers or splinters in wood pulp that may show up as imperfections in the finished sheet of paper made from it, if not previously removed. It is also found in groundwood pulp. Sometimes called slivers.

SHORT-DWELL COATER: A coating head developed primarily for thin papers to provide better runnability than conventional blade coaters by utilizing an optimum combination of forces on the blade, water transport from color to paper, and flow characteristics.

SHORTWOOD: Wood logs that have been cut into lengths suitable for use in pulp mills.

SIDE HILL SCREEN: A steeply inclined screening device commonly used in deinking systems. As pulp slides down the large screen deck by gravity, ink is drained off with the water.

SINGLE TIER DRYING: A dryer system arrangement in which the sheet passes over only a single row or bank of traditional dryer drums, rather than serpentine fashion over alternating top and bottom dryer drums. This reduces draw in the dryers and reportedly boosts overall drying efficiency.

SKIM: (1) The removal of the top layer or scum from the surface of a liquid. (2) The thin top or bottom layer of paper on a lined sheet of paper or board.

SLIVERS: See SHIVES.

SODA PROCESS: A chemical pulping process that consists of the reduction of chips to their individual fiber components by use of cooking liquor made up of caustic soda solution, the recovery and preparation of this liquor, or the treatment of pulp and paper produced from it.

SODA PULP: Pulp made by the cooking of chips from deciduous or broadleaf trees in a sodium hydroxide or caustic soda solution.

SODIUM HYPOCHLORITE (NaOCl): A chemical used as a bleaching agent in some multi-stage pulp mill bleach plants.

SODIUM SILICATE (Na2SiO3): A chemical used as a size material in papermaking to provide water and ink penetration resistance in the sheet. It is also used to make up an adhesive in corrugated paperboard and for the lamination of solid fiberboard. Commonly known as water glass.

SOFT NIP CALENDER: A type of on-machine calender that utilizes a roll with a soft, pliable, elastic cover that nips against a roll with a standard hard surface, to enhance gloss and general printability of printing and writing papers or boards. It is used in the manufacture of grades such as FILM COATED OFFSET.

SORTING: A manual or mechanical inspection operation carried on in the finishing room of a paper mill to remove imperfect sheets from an order. In a rag mill it refers to separating the different types of rags received for processing.

STICKIES: Sticking conditions occurring on paper machines using secondary recycled fiber containing materials such as ink, tars, latex, adhesives, and other organic compounds. Also refers to sticky materials called tackies and depositable pitch.

STOCK PREPARATION: The area of a paper mill where pulp is received from an on-site or off-site pulp mill, prepared for storage in slurry form, mechanically treated in beaters and refiners, mixed with other pulps, additives, dyes, and chemicals, and then cleaned and generally processed prior to sheet formation on the paper machine. Once known as the beater room.

STOCK: (1) The fibrous mixture in a paper mill that is ready to make into paper. It may consist of one or more types of beaten or refined pulps, with or without suitable fillers, dyes, additives, and other chemicals. Also called furnish and stuff. (2) Paper suitable for a particular use, such as coating raw stock, milk bottle stock, tag stock, towel stock, etc.

STONE GROUNDWOOD (SGW): Pulp made by abrading wood logs against a revolving stone, usually at atmospheric pressure but sometimes done under pressurized conditions.

SUBSTITUTION: A general reference to the use of chlorine dioxide as a substitute for elemental chlorine in the chlorination stage of a pulp bleaching sequence.

SULFATE: A general reference to the sulfate process and paper and board products manufactured from sulfate pulp.

SULFITE PROCESS: An acid pulp manufacturing process in which chips are reduced to their component parts by cooking in a pressurized vessel using a liquor composed of calcium, sodium, magnesium, or ammonia salts of sulfurous acid.

SULFITE PULP: Pulp made by the sulfite process.

SULFITE: A general reference to the sulfite process and paper and board products subsequently manufactured from sulfite pulp.

SULFUR: A yellow-colored, naturally occurring element that is burned in acid plants of sulfite pulp mills. It produces sulfur dioxide (SO2) which is used in making up calcium, sodium, magnesium, and ammonia base sulfite cooking liquors (sulfite process).

SUPERBATCH: A modified batch digester system (Sunds Defibrator) in which spent cooking liquors are collected in a series of accumulator vessels for use in subsequent batch cooks. See RAPID DISPLACEMENT HEATING (RDH).

SUPERCALENDER: An auxiliary piece of papermaking equipment used on some paper machines to obtain a denser paper with a higher finish than can be obtained on a conventional calender. It uses paper or cotton-covered rolls arranged alternately with metal rolls through which the sheet is passed. It is typically used off-line to produce supercalendered (SC) paper.

SURFACTANT: A material that enhances and stabilizes the ability of one liquid to disperse itself into another. Also known as emulsifying agent or emulsifier.

TCF PULP: Bleached pulp produced using a totally chlorine free bleaching process. See TOTALLY CHLORINE FREE.

TCF: See TOTALLY CHLORINE FREE.

THERMO-CHEMIMECHANICAL PULP (TCMP): See CHEMITHERMOMECHANICAL PULP (CTMP).

TOPWIRE FORMER: Former unit(s) normally located on top of a fourdrinier unit, in effect, adding a forming "side" to the top of the sheet.

TOTALLY CHLORINE-FREE: A reference to pulp bleaching sequences that do not use elemental chlorine or any chlorine containing compounds at all, including chlorine dioxide. Typically TCF processes employ hydrogen peroxide, oxygen, ozone, peracetic acid, etc., alone or in various combinations.

TRAILING BLADE COATER: See BLADE COATER.

TWINWIRE FORMER: A type of multi-ply paper or paperboard forming unit having two wires (or fabrics) between which the sheet is formed.

TWO-SIDEDNESS: The visual difference between the top or felt side of a sheet of paper and the bottom or wire side.

VACUUM BOX: A perforated, covered, narrow box extending across and over the area through which the paper machine wire (forming fabric) runs so that it will remove water from the forming paper web by the use of suction created by a pump connected to it. It is also used to remove water from a wet felt (press fabric) in the same manner. Sometimes referred to as a suction box and a uhle box.

VACUUM WASHER: Pulp mill equipment designed to separate undesirable components in a pulp slurry from the acceptable fibers using an evacuated, partially immersed, screen-covered cylinder to pick up the fibers from the slurry to form a mat, which is removed for further processing.

VELLUM: A high-quality parchment sheet made from young calves skins. It is used for making up deluxe books, certificates, diplomas, stationery, and other especially high-grade documents.

WASHING DEINKING: A process of removing ink from recycled furnishes using washing equipment rather than flotation deinking technology.

WASHING: The process of separating soluble, undesirable impurity components of pulp slurries from the fibers. Normally this is done after cooking, and during and after the bleaching operation by the use of some method of screening combined with the use of fresh water and other liquids.

WASTEPAPER: All types of used paper or paper (discarded or not considered fit for a particular use) that provides a source of fiber for the manufacture of some papers, paperboards, and chipboards. Wastepaper is usually processed in a separate plant section of a mill that includes pulping/deinking (but not cooking), screening, cleaning, etc. Also known as old paper or recovered paper.

WATERMARK DANDY ROLL: A wire-covered skeleton roll covered with a woven wire cloth and located on the top of a paper machine wet end wire (forming fabric). It has a raised surface pattern that impresses the wet web so that the design appears in the final dried sheet. See also WATERMARKS and DANDY ROLL.

WATERMARKS: Translucent marks or designs made in paper as the result of modifying the fiber orientation of the wet web. It is done by impressing the sheet with raised patterns on rolls so that the localized, displaced fibers produce more transparent areas in the final dried sheet. They are usually used for identification purposes. Also known simply as marks. See also WATERMARK DANDY ROLL.

WEB: A continuous sheet of paper produced and rolled up at full width on the paper machine. It is not cut into sheets, but is used directly from rolls for converting into consumer goods.

WET END: The section of the head end of a paper machine, which includes the headbox, wire (forming fabric), and wet press sections, where the sheet is formed from the stock furnish and most of the water is removed before entering the dryer section. Sometimes called wire end.

WHITE LIQUOR CLARIFICATION: The removal of calcium carbonate (CaCO3) and other impurities from the causticizing liquor, usually by gravity sedimentation in a unit called a clarifier. This takes place in the liquor recausticizing process of a pulp mill to obtain a clear liquor for cooking wood.

WHITE LIQUOR: Cooking liquor formed by refortifying green liquor in the causticizing operation of an alkaline-type pulp mill so that it contains the active chemicals that will reduce chips into their fiber components by dissolving the lignin cementing material during the digester operation, thereby producing pulp.

WIRE: A general reference to the forming fabric used on the wet end of a paper machine. These continuous screen belts, on which fibers in the pulp slurry form a sheet of paper as water drains away, are commonly made of woven plastic fibers (or monofilament strands), whereas they once were generally made of woven metal wires, such as bronze (some machine still use metal wires). The term may also be used to generally indicate the forming section of a fourdrinier type of paper machine. See also FOURDRINIER WIRE.

# About the Author

Ken L. Patrick is president of Paper Industry Communications, Inc., in Atlanta, Ga. He served as editor of *Pulp & Paper* magazine for almost 20 years, from 1977 to 1996. Prior to joining the editorial staff of the magazine, he worked in the paper industry for five years as a technical communications supervisor at International Paper Co.'s Erling Riis Research Laboratory in Mobile, Ala., and the Natchez, Miss., mill.